T0137261

Lecture Notes in Computational Science and Engineering

124

Editors:

More information about this series at http://www.springer.com/series/3527

Michael Schäfer • Marek Behr • Miriam Mehl •
Barbara Wohlmuth
Editors

Recent Advances in Computational Engineering

Proceedings of the 4th International Conference on Computational Engineering (ICCE 2017) in Darmstadt

Editors
Michael Schäfer
Graduate School of Computational Engineering
Technische Universität Darmstadt
Darmstadt, Germany

Marek Behr
Lehrstuhl für CATS
RWTH Aachen University
Aachen
Germany

Miriam Mehl
Institut für Parallele und Verteilte Systeme (IPVS)
Universität Stuttgart
Stuttgart, Germany

Barbara Wohlmuth
Lehrstuhl für Numerische Mathematik
Technische Universität München
Garching bei München, Germany

ISSN 1439-7358 ISSN 2197-7100 (electronic)
Lecture Notes in Computational Science and Engineering
ISBN 978-3-030-06738-0 ISBN 978-3-319-93891-2 (eBook)
https://doi.org/10.1007/978-3-319-93891-2

Mathematics Subject Classification (2010): 49-06, 60-06, 65-06, 68-06, 70-06, 74-06, 76-06, 80-06

This Springer imprint is published by the registered company Springer Nature Switzerland AG
The registered company address is: Gewerbestrasse 11, 6330 Cham, Switzerland

Preface

Methods of computational engineering have become a key technology in all fields of engineering sciences during the last decades. Computer-based modeling, analysis, simulation, and optimization offer attractive, cost-efficient ways to investigate increasingly complex engineering applications and to develop new technical solutions. Because of this, computational engineering contributes to major issues of technical developments in economy and society in areas such as energy, health, safety, and mobility.

This volume is a collection of selected contributions from the 4th International Conference on Computational Engineering that took place in Darmstadt (September 2017). The event was a continuation of a series of conferences—the previous ones were held in Herrsching (October 2009), Darmstadt (October 2011), and Stuttgart (October 2014). The conference brought together researchers working on computational methods in all disciplines of engineering, applied mathematics, and computer science. The aims were to discuss the state of the art in this challenging field, exchange experiences, develop promising perspectives for future research, and initiate further cooperation.

Besides general aspects of numerical concepts and modeling techniques, special focus was put on multiphysics coupling, discontinuous Galerkin methods, reactive and multiphase flows, uncertainty quantification, and high-performance computing techniques. Altogether, the book contains a collection of different types of problem-driven and methodology-driven approaches in the field of computational engineering. We hope that it provides the reader with a valuable source of inspiration for his/her own work.

Darmstadt, Germany
March 2018

Michael Schäfer

Organization

The 4th International Conference on Computational Engineering (ICCE 2017) is organized by the Graduate School of Computational Engineering (GSC CE) at Technische Universität Darmstadt in cooperation with the Aachen Institute for Advanced Study in Computational Engineering Science (AICES) at RWTH Aachen University, the International Graduate School of Science and Engineering (IGSSE) at the Technical University of Munich, and the Stuttgart Research Center for Simulation Technology (SimTech) at the University of Stuttgart.

Chairpersons

Marek Behr	RWTH Aachen University
Miriam Mehl	Universität Stuttgart
Michael Schäfer	Technische Universität Darmstadt
Barbara Wohlmuth	Technische Universität München

Contents

Contributors

Harish Abubaker Lehrstuhl für Strömungsmechanik, Friedrich-Alexander-Universität Erlangen-Nürnberg, Erlangen, Germany

Benjamin Berkels AICES Graduate School, RWTH Aachen University, Aachen, Germany

Johanna Biehl Technische Universität Darmstadt, Darmstadt, Germany

Andre Boye Fraunhofer IVV Dresden, Dresden, Germany

Antonio Delgado Friedrich-Alexander Universität Erlangen-Nürnberg, Lehrstuhl für Strömungsmechanik, Erlangen, Germany

Xin Huang Graduate School of Computational Engineering, Technische Universität Darmstadt, Darmstadt, Germany

Institute of Numerical Methods in Mechanical Engineering, Technische Universität Darmstadt, Darmstadt, Germany

Fabian Kayatz Fraunhofer-Institut für Verfahrenstechnik und Verpackung IVV, Dresden, Germany

Anne Kikker Technische Universität Darmstadt, Darmstadt, Germany

Graduate School CE, Technische Universität Darmstadt, Darmstadt, Germany

Julia Kowalski AICES Graduate School, RWTH Aachen University, Aachen, Germany

Florian Kummer Technische Universität Darmstadt, Darmstadt, Germany

Jens Lang Graduate School of Computational Engineering, Technische Universität Darmstadt, Darmstadt, Germany

Department of Mathematics, Technische Universität Darmstadt, Darmstadt, Germany

Jens-Peter Majschak Fraunhofer-Institut für Verfahrenstechnik und Verpackung IVV, Dresden, Germany

Bram Metsch Fraunhofer SCAI, Sankt Augustin, Germany

Christopher Müller Graduate School of Computational Engineering, Technische Universität Darmstadt, Darmstadt, Germany

Department of Mathematics, Technische Universität Darmstadt, Darmstadt, Germany

Manuel Münsch Lehrstuhl für Strömungsmechanik, Friedrich-Alexander-Universität Erlangen-Nürnberg, Erlangen, Germany

Roman Murcek Fraunhofer IVV Dresden, Dresden, Germany

Fabian Nick Fraunhofer SCAI, Sankt Augustin, Germany

Hans-Joachim Plum Fraunhofer SCAI, Sankt Augustin, Germany

Michael Schäfer Graduate School of Computational Engineering, Technische Universität Darmstadt, Darmstadt, Germany

Andreas Schmitt TU Darmstadt, Graduate School of Computational Engineering, Darmstadt, Germany

TU Darmstadt, Institute of Numerical Methods in Mechanical Engineering, Darmstadt, Germany

Martin Schreiber University of Exeter, Mathematics/Computer Science, Exeter, UK

Kai Schüller AICES Graduate School, RWTH Aachen University, Aachen, Germany

Anand Sivaram Lehrstuhl für Strömungsmechanik, Friedrich-Alexander-Universität Erlangen-Nürnberg, Erlangen, Germany

Christopher Spannring Graduate School of Computational Engineering, Technische Universität Darmstadt, Darmstadt, Germany

Department of Mathematics, Technische Universität Darmstadt, Darmstadt, Germany

Stefan Ulbrich Technische Universität Darmstadt, Darmstadt, Germany

Sebastian Ullmann Graduate School of Computational Engineering, Technische Universität Darmstadt, Darmstadt, Germany

Department of Mathematics, Technische Universität Darmstadt, Darmstadt, Germany

Simon Wagner Friedrich-Alexander Universität Erlangen-Nürnberg, Lehrstuhl für Strömungsmechanik, Erlangen, Germany

Alexander G. Zimmerman AICES Graduate School, RWTH Aachen University, Aachen, Germany

Optimization of Design Parameters of CIP Spray Cleaning Nozzle Using Genetic Algorithm

Harish Abubaker, Anand Sivaram, Manuel Münsch, Roman Murcek, Andre Boye, and Antonio Delgado

Abstract The spray cleaning of surfaces is a standard task in the food and pharmaceutical industries. At present, the development of such nozzles is based on semi-empirical methods, experience and iterative prototyping. This almost makes it prohibitive to develop nozzles for specific customer requirements due to higher time and cost. A Virtual Engineering approach to design and optimize unlimited number of nozzle designs can overcome this.

In this work, a parametric study is carried out to recognize design parameters that have maximum impact on flow. Based on this, a Multiobjective optimization code based on Genetic Algorithm is developed to optimize the design parameters of a full cone nozzle. CFD Simulations were used to estimate the objective functions.

In future, the work shall be extended by comparing genetic algorithm with other optimization algorithms and replacing expensive CFD simulations with meta-models.

Keywords Multi-objective optimization · CFD · Genetic algorithm · Full cone nozzle

---

H. Abubaker (✉) · A. Sivaram · M. Münsch · A. Delgado
Lehrstuhl für Strömungsmechanik, Friedrich-Alexander-Universität Erlangen-Nürnberg, Erlangen, Germany
e-mail: harish.abubaker@fau.de; https://www.lstm.uni-erlangen.de/

R. Murcek · A. Boye
Fraunhofer IVV Dresden, Dresden, Germany

M. Schäfer et al. (eds.), *Recent Advances in Computational Engineering*, Lecture Notes in Computational Science and Engineering 124,
https://doi.org/10.1007/978-3-319-93891-2_1

1 Introduction

Hygiene is of prime importance in industries like food and pharmaceutical industries. For this purpose, spray nozzles operating in the range 1–6 bar are preferred because of the risk of aerosol formation and also the price for running and setting up of such nozzles are minimal. Optimization of such nozzles would result in considerable savings due to reduction in time required for cleaning and the resources consumed. Also with the rising concerns for environment, the reduction of waste water from cleaning also possess a challenge. Though the costs vary depending on the industry, an estimate shows that freshwater costs range from 0.19 to 2.30 €/m^3 in the brewery industry [1], up to 4 €/m^3 in the pharmaceutical industry [2]. Waste water treatment of water with presence of hormones or carcinogenic substances can cost as high as 300 €/m^3 [3].

An additional issue is the development of new nozzle designs for specific purposes. Traditionally, the development of such nozzles is based on semi-empirical methods, experience of the designer and iterative prototyping. This approach results in longer time for development as well as high costs for the trial technical testing. A virtual engineering approach in design and optimization can overcome these shortcomings by enabling the early analysis of unlimited number of nozzles virtually and at reasonable costs.

In this work, an optimization algorithm is developed which can find optimum design parameters based on the objective function provided for optimization. The objective functions are evaluated using CFD simulations. Also, an effort has been taken towards developing a meta model for replacing the computationally expensive CFD simulations. For this, a neural network was developed and trained to predict the objective functions.

2 Simulation Model

Figure 1 shows the nomenclature of a generic full cone nozzle. The water coming in passes through vanes which have helical form. This increases the tangential velocity component of the incoming stream. This produces a swirling flow inside the swirl chamber. Due to the high centrifugal forces, the streams are pushed towards the edges and produces a cone angle as the water stream exits through the orifice. Full cone nozzles can produce uniform liquid distribution over the entire circular impact area. Among different nozzle types, full cone nozzles are known to produce the largest drop sizes. The velocity of the droplets depend on the droplet size. Smaller droplets will have a higher velocity initially, but gets reduced faster. Larger droplets can travel for longer distance without much velocity reduction. These advantages of full cone nozzle make it a good choice for cleaning purposes.

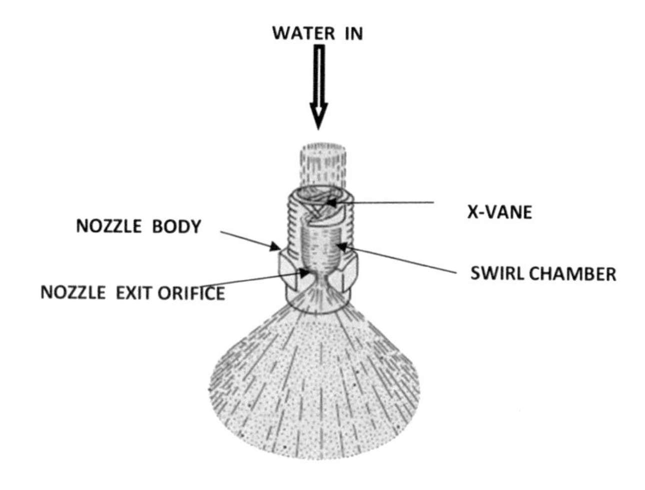

Fig. 1 Full cone nozzle nomenclature [4]

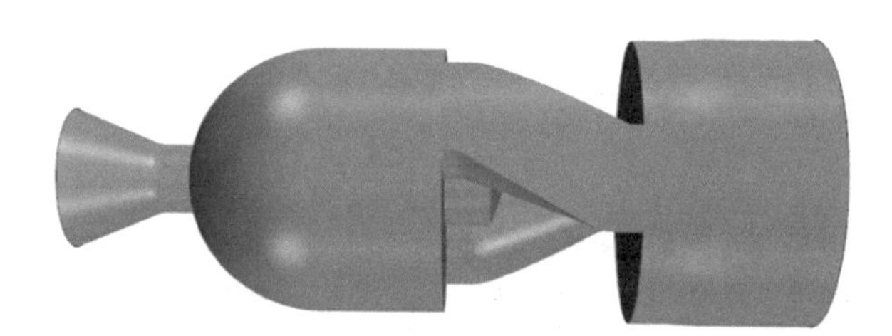

Fig. 2 Fluid volume of nozzle

2.1 *Nozzle Geometry*

The simulation model considered was based on a patent from the company Lechler GmbH [5]. Figure 2 shows the fluid volume of the nozzle. The model considered has two helical ducts and a blind hole is provided at the base of swirl chamber to reduce pressure gradient, thus preventing air entrainment which will produce a hollow cone. The initial parametric study have been done using this model and for optimization, a reduced model shown in Fig. 3 was considered. This was done to reduce the computational effort and can be considered sufficient to understand the feasibility of using evolutionary algorithms in nozzle design optimization.

As the optimization objectives were to obtain uniform flow at the nozzle exit, four parameters near to the nozzle exit were considered for optimization, as can be seen on the right of Fig. 3:

- Exit diameter, R
- Angle at exit, Angle
- Orifice length, l
- Orifice diameter, r

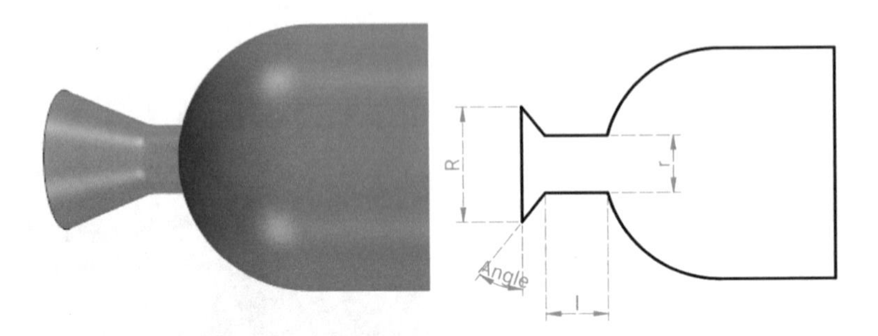

Fig. 3 Fluid volume considered for optimization

2.2 *Boundary Conditions*

For boundary conditions, no slip wall was specified at the nozzle walls. At the nozzle exit, pressure boundary condition specified at atmospheric pressure was used. For the inlet, simulation of the model in Fig. 2 was done and the flow condition at the plane of swirl chamber entrance was captured and this was used and mapped onto the inlet plane as the inlet condition. This ensured that the swirling flow streams were produced without having the computational efforts for simulating the helical ducts. The results at the nozzle exit of the full model was compared with that of reduced model and was found to be matching.

The working fluid considered was water at 25 °C with density of 997 kg/m^3 and dynamic viscosity of 8.89×10^{-4} Pa s. The Reynolds number based on the nozzle inlet diameter and inlet velocity is in the order of 10^4.

3 Methodology

3.1 *Governing Equations*

The flow simulations are computed for a three-dimensional, unsteady and turbulent flow. The working fluid was assumed to be incompressible and having constant material properties. The governing equations for the fluid flow in Einstein's notation are the following:

Continuity equation:

$$\frac{\partial u_i}{\partial x_i} = 0 \tag{1}$$

Momentum equation:

$$\rho\left[\frac{\partial u_i}{\partial t}+\frac{\partial u_j u_i}{\partial x_j}\right]=-\frac{\partial p}{\partial x_i}-\frac{\partial \tau_{ij}}{\partial x_j}+\rho g_i \quad (i=1,2,3) \tag{2}$$

where ρ is density, u_i is the cartesian velocity component, g_i the forces and τ_{ij} the stress tensor.

3.2 Numerical Settings

The numerical simulation was carried out using the finite volume CFD code StarCCM+. A second order upwind scheme was used for space discretization. Due to the unsteady nature of flow, an implicit first order discretization scheme was chosen for the time discretization. The time step size was chosen to be 10^{-4} s for ensuring CFL number in the order of 1. A segregated flow solver (SIMPLE) was used which solves the governing equations of mass and momentum in a sequential manner.

The meshing was done using unstructured polyhedral mesh with a base size in order of 10^{-4} m. Prism cells were used to refine the regions near the nozzle walls. As the meshing needed to be done automatically for different designs, the base size was specified as function of the orifice radius. This resulted in meshes with size in order of 0.2 million.

For modelling turbulence, Reynolds stress turbulence model was used. This model is able to provide better predictions [6] for turbulent flows that have high degree of anisotropy, rotation effects and zones of recirculation.

The mean velocity at the nozzle exit was monitored and when this has reached an asymptotic limit, the simulation was assumed to be converged and stopped.

3.3 Objective Functions

The different designs were compared based on two parameters at the nozzle exit:

- Average flow velocity at exit
- Coefficient of variation (CoV) of velocity at exit

where CoV is the ratio of standard deviation to the mean $\left(\frac{\sigma}{\mu}\right)$.

CoV can be considered a good parameter to judge the uniformity of flow at exit. Having a low CoV means the velocity profile would be more uniform. A jet that has uniform flow at exit would be more stable [7]. This is due to the fact that, when a flow exits the nozzle, the effect of wall no longer exists. This causes redistribution of momentum inside the jet, creating radial velocity components. This would result in the spreading and faster breakup of jet. When the jet has a profile like a top hat profile, this momentum redistribution can be kept to a minimum and hence also the jet spread.

Since the simulation was not involving the atmosphere into which jet emerges, it is not possible to prescribe exact objective functions that would aid in cleaning action.

3.4 Optimization Methods

In the previous section, the two objective functions were defined. If only the minimization of CoV was considered, the resulting nozzle would resemble a pipe like device which tries to minimize any disturbance inside and tries to achieve a fully developed flow. But, this would then minimize the maximum velocity at the nozzle exit. For fixed mass flow, it is better to have a nozzle that produces maximum velocity at the exit. Presence of these conflicting objectives give rise to a multiobjective optimization problem.

For solving a multiobjective optimization problem, an efficient method is using the concept of pareto optimality. In this concept, one design is considered to be dominating if it is better than all other designs in one objective and better or equal in other objectives. A set of pareto optimal solutions is known as a pareto front. This depicts all the non-dominated designs. One objective of a design from the front cannot be improved without degrading another objective. This results in a set of designs that are equally good and the designer can select based on further considerations.

Many methods can be used to find the pareto front. From the initial parametric study (Fig. 4), it was seen that all the geometric parameters have non-trivial influence on the flow condition at exit, this results in a large design space for optimization. In the figure, impact factor η is the ratio of percentage change in objective function to percentage change in geometric parameter with respect to a reference. The parameters relevant for this work are explained below:

- D_o—contraction diameter
- L_o—contraction length
- α—angle of expansion at exit

For such cases, stochastic approaches like evolutionary algorithms are a better choice. In this work, genetic algorithm (GA) was chosen as the optimization algorithm. GA is based on the evolutionary idea of survival of the fittest. A flow chart of GA is shown in Fig. 5. GA starts by creating a random population. The population consists of individuals (chromosomes) which has encoded values (genes) which can describe the individual.

The code was written in Python using the library DEAP [8] (Distributed Evolutionary Algorithms in Python). For encoding the real values of geometric parameters into gene values, a linear mapping was utilized. All the parametric value

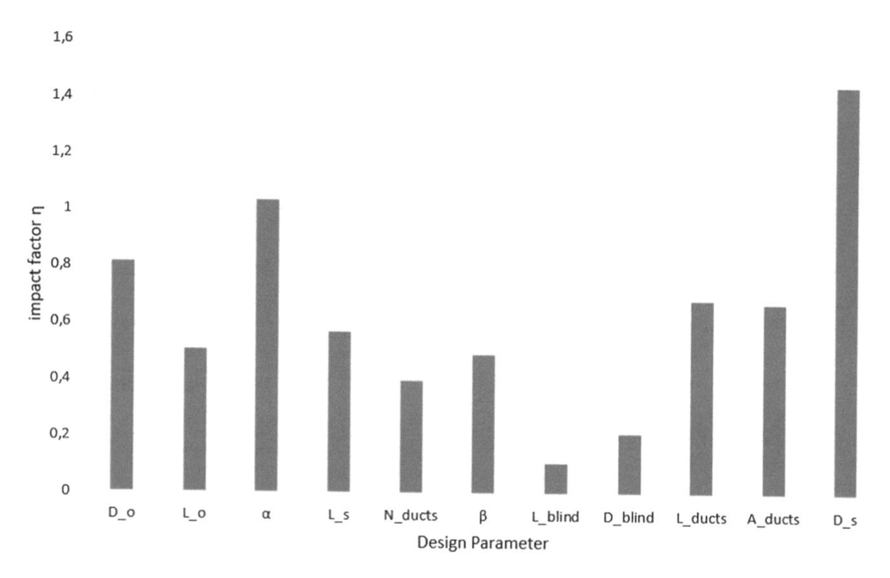

Fig. 4 Impact factor η for CoV

was mapped into range 1–100 and hence the gene values always have the same range of values. This helps to implement genetic operators that operate on all gene values irrespective of the actual value differences.

Parents are selected from the generation based on their fitness values. These parents are then used to produce new offspring for the next generation through the crossover and mutation operations. The process continues till a specified convergence criterion is satisfied. In this work, it was limited to 20 generation as no change was observed in the pareto front after the 15th generation.

The operations enclosed in the box at right in Fig. 5 are for fitness evaluation. The whole process was automated using macros in StarCCM+. The simulation can read the gene value from chromosome and generate corresponding CAD. This is then meshed and simulation is run. Once the convergence criterion are satisfied, the required objective functions are extracted and fitness of the design can be evaluated.

For selection of parents, two methods were employed, NSGA2 [9] and SPEA2 [10]. The details of which are explained in detail in the references cited. Once the parents are selected, the genetic operators, crossover and mutation are done to obtain next generation. The crossover operation mates two parents to produce two offspring. The function implemented in this work can be seen in Fig. 6. A random position is selected and the genes before and after this crossover point are exchanged between the parents.

Figure 7 shows the mutation operation. A position is selected at random and then the gene at this position is replaced by a random generated gene value. The mutation process is necessary as it helps to keep the diversity in population and prevents the algorithm getting stuck in any local optima.

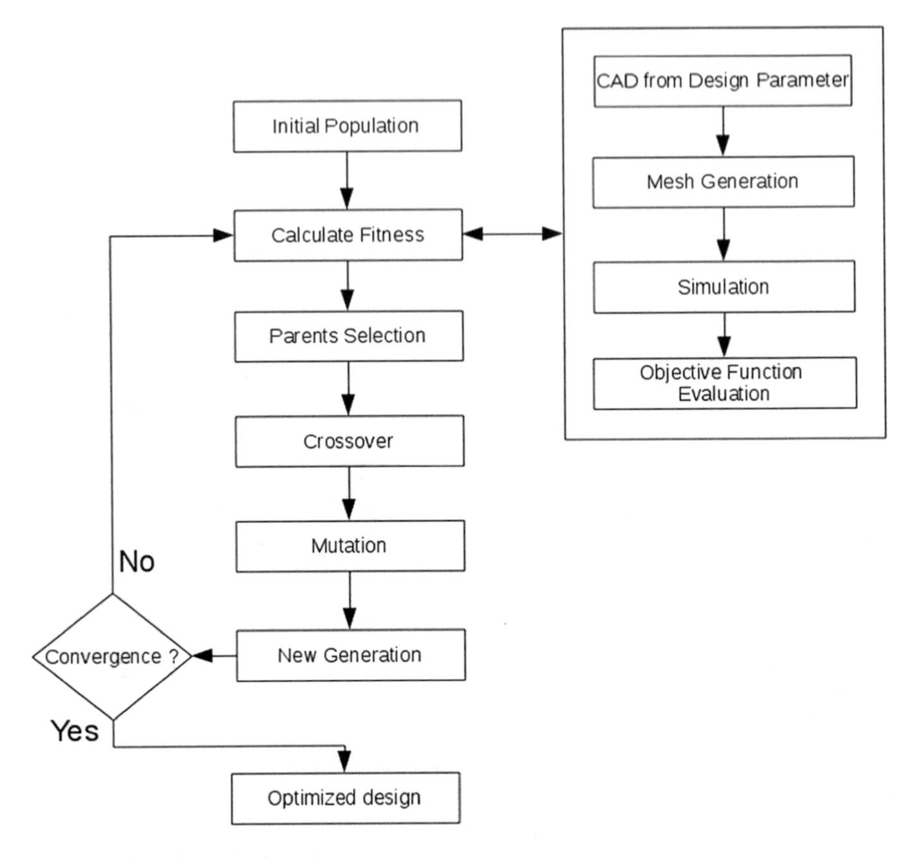

Fig. 5 Genetic algorithm flow chart

The code was run on the Emmy Cluster at RRZE, Erlangen using 200 processors. The simulations were distributed equally to all processors in each generation based on the number of designs that needed to be evaluated.

3.5 *Meta-models*

The major time consumed during the algorithm run was for CFD simulations. For replacing the computationally intensive simulations using meta models which can predict the objectives, a neural network was developed in MATLAB. The schematic of network is shown in Fig. 8. It consisted of four input neurons corresponding to the four design parameters and two output neurons, one for each of the objective functions. The hidden layer consisted of 15 neurons.

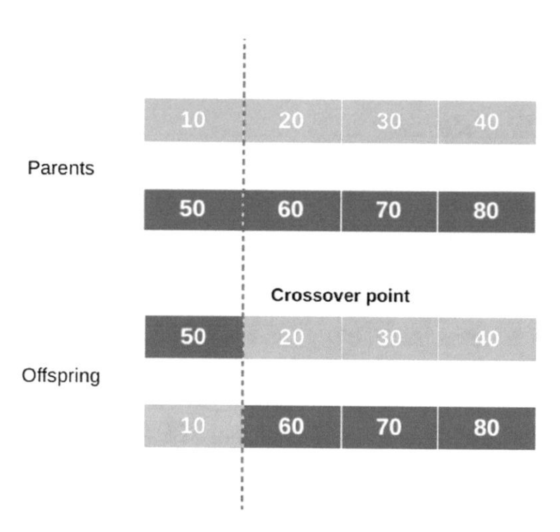

Fig. 6 Crossover function

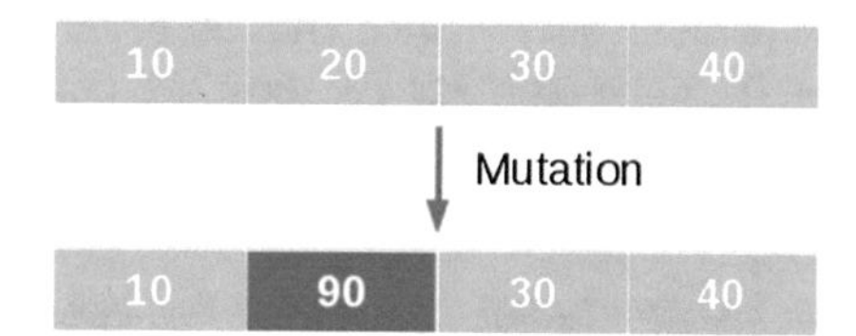

Fig. 7 Mutation function

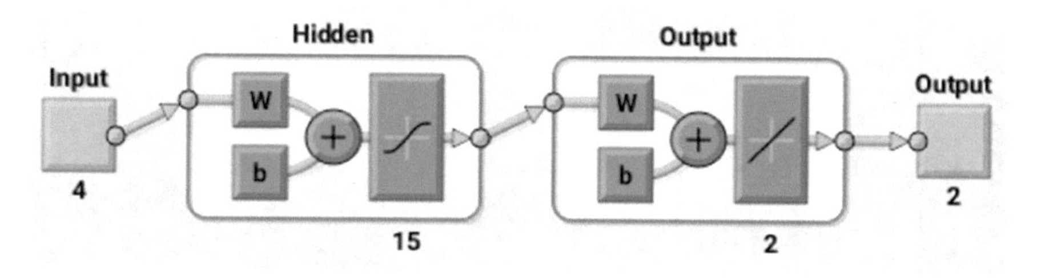

Fig. 8 Neural network schematic

In the actual optimization algorithm, this network was not used as the training data required had to be generated during the optimization run. The network was created as possible future replacement of CFD simulations.

4 Results

4.1 Optimization Results

The genetic algorithm was run for 20 generations and the results are discussed here. There was no change observed in the pareto front from generation 15 to 20 and hence code was not run further. Figure 9 shows the variation of average velocity of exit with generation. The individual with the highest velocity in each generation is considered. It can be seen that the a good individual had been identified early on and was retained in subsequent generations. The figure also shows the comparison between NSGA2 and SPEA2 selection algorithms. In this case, NSGA2 was able to produce better designs compared to SPEA2, but SPEA2 was able to improve on the design as generations evolved.

Similarly, the variation of coefficient of variation versus generation has been plotted in Fig. 10. The individual with lowest coefficient of variation value had been considered. Here also NSGA2 selection procedure fared better in reducing the coefficient of variation.

Figure 11 shows the pareto front obtained after 20 generations. It shows all the non dominated designs. It can be seen that the most designs on the front have low coefficient of variation with higher velocity at nozzle exit. Each of this design are

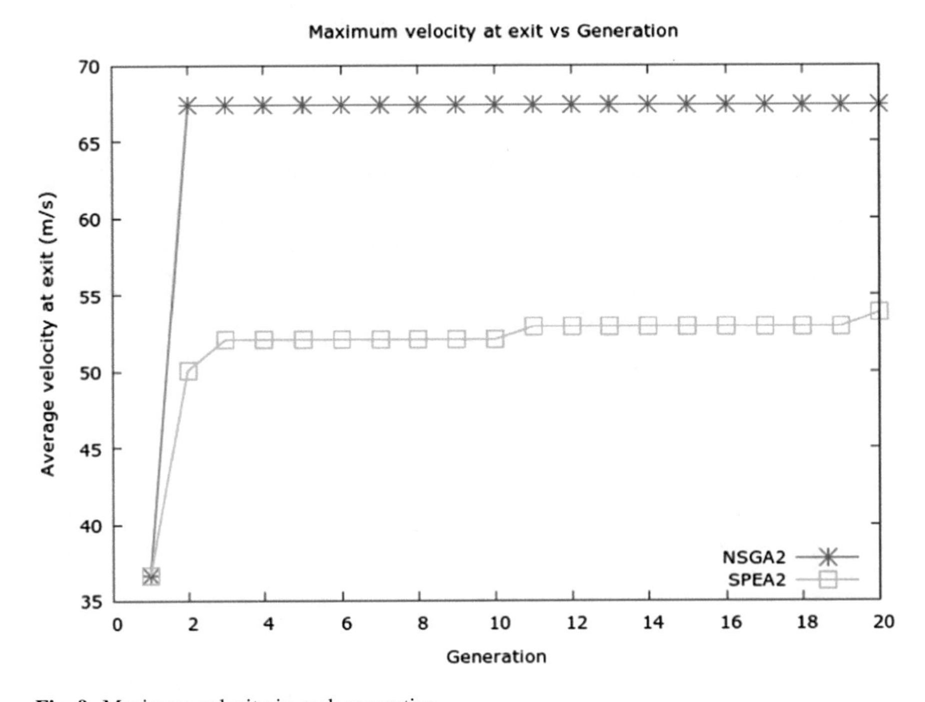

Fig. 9 Maximum velocity in each generation

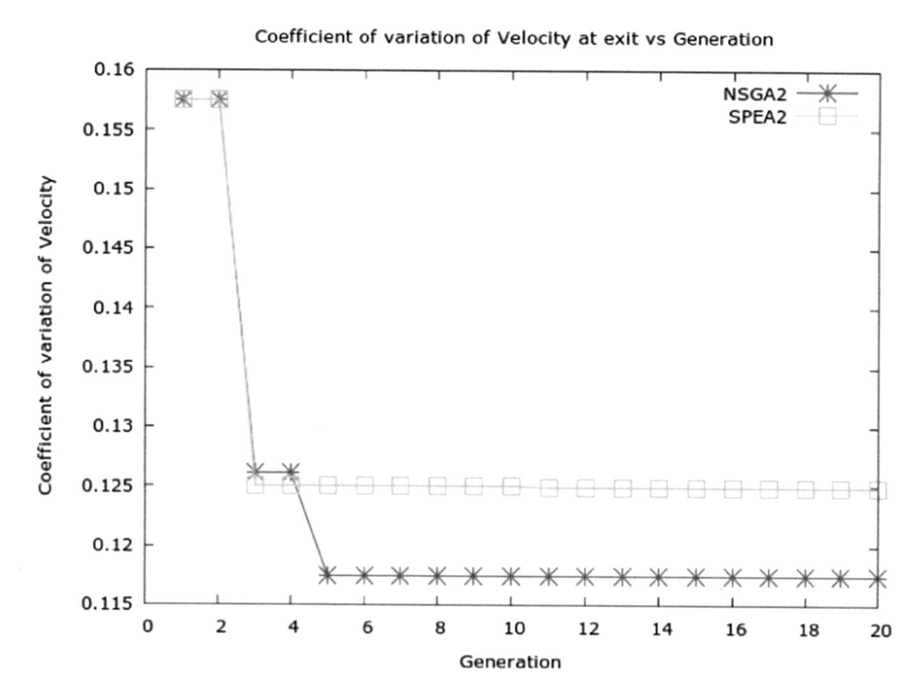

Fig. 10 Minimum coefficient of variation in each generation

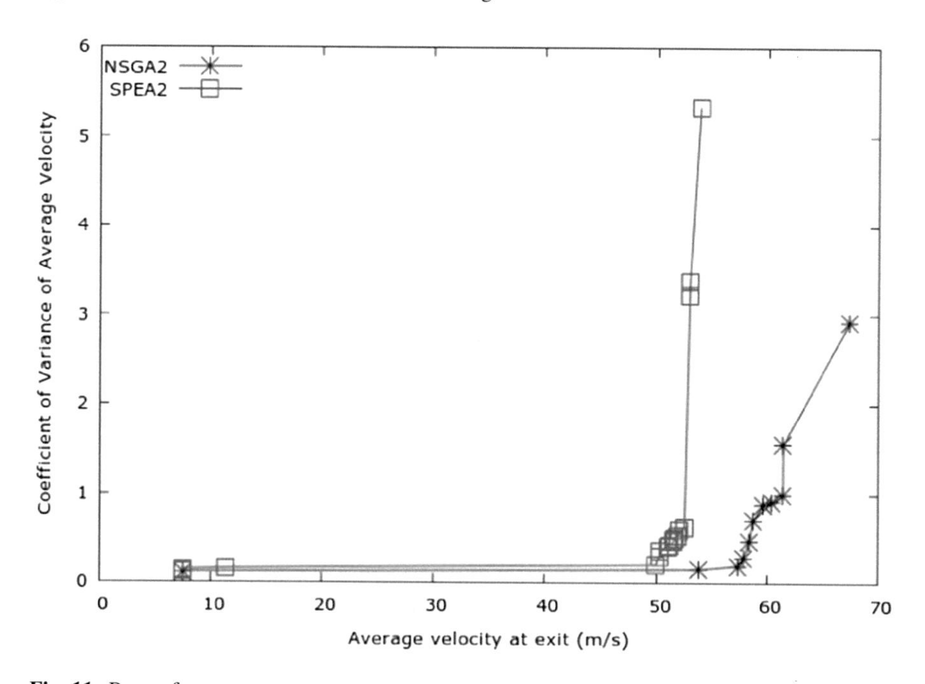

Fig. 11 Pareto front

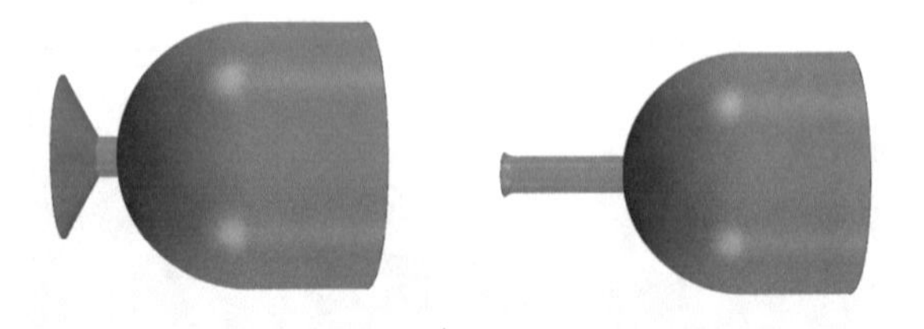

Fig. 12 Design from initial and final generation

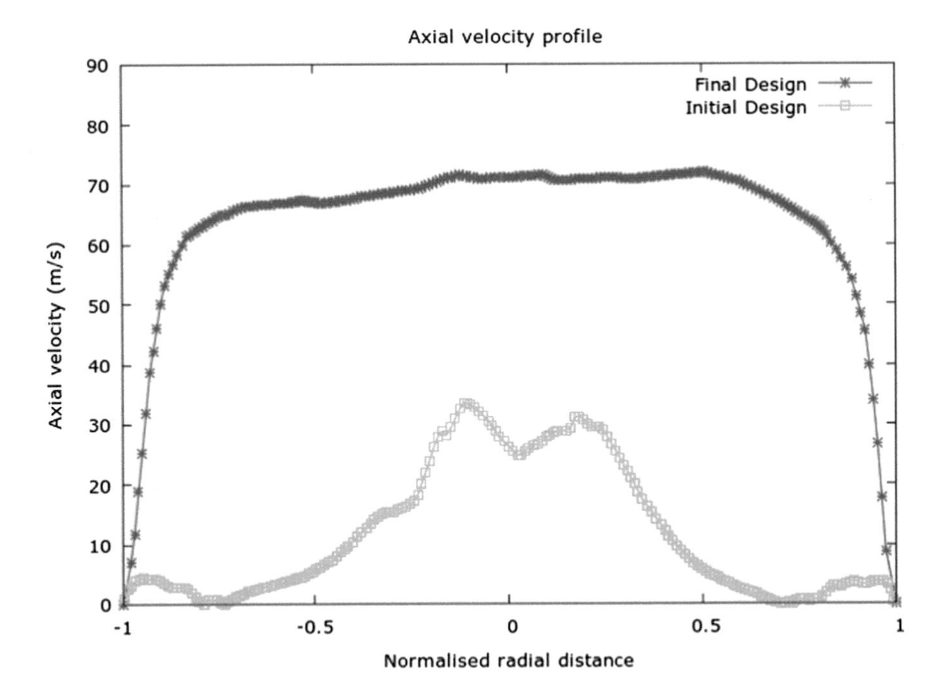

Fig. 13 Velocity profile comparison

equally good and the choice would be depending on the designer. A higher velocity design can be opted by sacrificing on the coefficient of variation or vice versa.

Two designs, one from initial generation and second one from final generation are shown in Fig. 12. The corresponding velocity profiles at the nozzle exit are given in Fig. 13. The horizontal axis shows normalized radial distance so that both nozzles can be compared on similar scale. The radial distance is normalized with the nozzle exit radius. The average velocity have been almost doubled and from the figure it can be seen that the profile has become more uniform, i.e., the coefficient of variation has been lowered. The algorithm have increased the contraction length which help to produce a fully developed flow which is uniform and by making the exit radius close to contraction radius, any chances of disturbance occurring at this

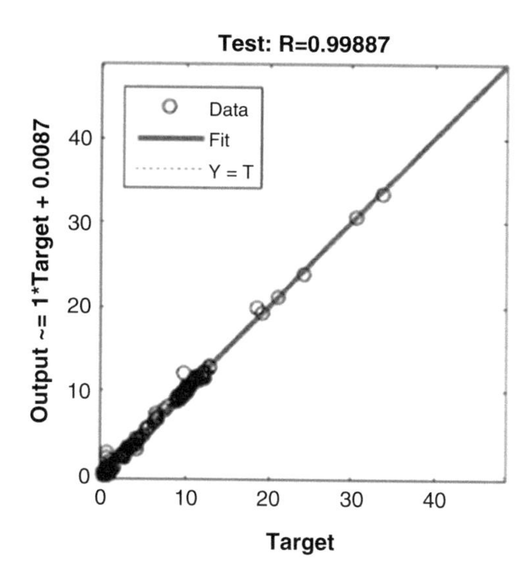

Fig. 14 Neural network test results

stage are minimized. The contraction radius have been brought close to the lower limit specified, which results in producing a higher velocity.

4.2 Neural Network Results

The neural network shown in Fig. 8 was able to predict the velocity at exit with good accuracy. But the network was not able to predict the coefficient of variation accurately. Figure 14 shows the plot of predicted versus actual velocities. Most of the points are found to be lying on the 45 degree line, indicating a good match between prediction and actual values.

5 Conclusion and Future Works

In this work, a multiobjective genetic algorithm has been developed to optimize the geometric parameters of a full cone nozzle close to the exit to produce a uniform flow at the exit. It was seen that genetic algorithm was able to optimize the nozzle design in accordance to the objective functions specified. For obtaining a uniform flow with higher velocity, the algorithm have evolved a nozzle with longer contraction length and nozzle exit diameter close to contraction radius. This shows that by specifying an objective function as per the requirement of the designer, optimized set of nozzles can be produced in lesser time and cost. A neural network was also trained which can be used to replace the computationally expensive CFD simulations. The predictions for velocity was found to be good, though the network was not able to predict coefficient of variation accurately. This can be overcome in future by increasing the number of hidden layers. Due to time constraint, this was not implemented in the present work.

For future works, multiphase simulation for simulating the jet flow after the nozzle exit needs to be created and objective function can be defined on the flow parameters at some point downstream from the nozzle exit. With such objective function, the exact cleaning process can be represented and optimized designs can be produced based on more accurate flow physics. A neural network or another meta model can be trained beforehand using simulation data and can be used in the optimization algorithm to replace simulations.

Acknowledgements The authors are thankful to Mr. Alexander Frühbeis for his contribution to the parametric study conducted as part of his master thesis.

References

1. Hien, O., Küpferling, E., Guggeis, H.: Wassermanagement in der Getränkeindustrie. BRAUWELT **23**, 640–643 (2008)
2. Graf, C.: Energieeffiziente Herstellung von Pharmawasser. Die Pharmazeutische Industrie **72**, 1797–1803 (2010)
3. Eckert, V., Bensmann, H., Zegenhagen, F., Weckenmann, J., Sörensen, M.: Elimination of hormones in pharamaceutical waste water. Die Pharmazeutische Industrie **74**, 487–492 (2012)
4. Jain, M., John, B., Iyer, K., Prabhu, S.: Characterization of the full cone pressure swirl spray nozzles for the nuclear reactor containment spray system. Nucl. Eng. Des. **273**, 131–142 (2014)
5. Schneider, M.: Vollkegeldüse. Vollkegeldüse, Lechler GmbH, 72555, Metzingen, DE, patent no. DE 10 2011 078 508 A1, 2011
6. Pope, S.B.: Turbulent Flows. Cambridge University Press, Cambridge (2000)
7. Birouk, M., Lekic, N.: Liquid jet breakup in quiescent atmosphere: a review. Atomization Sprays **19**(6), 501–528 (2009)
8. Fortin, F., De Rainville, F., Gardner, M., Parizeau, M., Gagné, C.: DEAP: evolutionary algorithms made easy. J. Mach. Learn. Res. **13**(1), 2171–2175 (2012)
9. Deb, K., Pratap, A., Agarwal, S., Meyarivan, T.A.M.T.: A fast and elitist multiobjective genetic algorithm: NSGA-II. IEEE Trans. Evol. Comput. **6**(2), 182–197 (2002)
10. Zitzler, E., Laumanns, M., Thiele, L.: SPEA2: improving the strength pareto evolutionary algorithm for multiobjective optimization. In: Evolutionary Methods for Design, Optimization and Control with Applications to Industrial Problems (2001)

Multilevel Optimization of Fluid-Structure Interaction Based on Reduced Order Models

Johanna Biehl and Stefan Ulbrich

Abstract We consider the optimal control of fluid-structure interaction (FSI) problems. In order to apply a multilevel optimization algorithm, we introduce a reduced order model using proper orthogonal decomposition (POD). The stability of the resulting system and the existence and uniqueness of solutions is ensured by enriching the reduced basis by suitable supremizers. Further, we reduce the coupling condition by reinterpretation of the variables as boundary integrals. Based on this reduced order model we describe a multilevel optimization algorithm which solves optimal control problems using a sequential quadratic programming (SQP) algorithm and show results for a benchmark problem.

Keywords Reduced order models · Proper orthogonal decomposition · Fluid-structure interaction · Multilevel optimization

1 Introduction and Outline

The interaction of fluid flows with elastic materials is part of many applications, for example in medicine, civil engineering or aerodynamics. Therefore it is of interest to control the flow and the corresponding deformation of the structure. Since the finite element discretization of this kind of problems gives rise to high dimensional nonlinear problems, the solution is computationally expensive. Moreover during an optimization process we have to calculate the state several times for different controls. Therefore we present a reduced order model based on proper orthogonal decomposition which is of much lower dimension. The idea is then to run the optimization algorithm on this approximation of the finite element discretization and use the full model only to control the quality of the reduced order model. That means, we estimate the error and calculate new snapshots on the finite element level.

J. Biehl (✉) · S. Ulbrich
Technische Universität Darmstadt, Darmstadt, Germany
e-mail: biehl@mathematik.tu-darmstadt.de

M. Schäfer et al. (eds.), *Recent Advances in Computational Engineering*, Lecture Notes in Computational Science and Engineering 124,
https://doi.org/10.1007/978-3-319-93891-2_2

The optimization itself, the calculation of descent directions and step lengths, is done on the reduced order model in order to save time and memory.

The paper is structured as follows. In Sect. 2 we introduce the notations and the spaces we need for modeling fluid-structure interaction (FSI) problems. Afterwards, in Sect. 3 we state the problem with help of the Arbitrary Lagrangian-Eulerian method. The discretization via Crank-Nicolson and finite element schemes is shown in Sect. 4. The proper orthogonal decomposition method is introduced in Sect. 5 and in Sect. 6 we introduce the model reduction in detail for FSI problems. In Sect. 7 we show how supremizers are used to ensure existence and stability of solutions and in Sect. 8 we state the multilevel optimization algorithm. Finally, in Sect. 9 we show numerical results.

2 Notation and Spaces

In this section we introduce the notation used to describe FSI problems. We frequently use the subscripts s or f to indicate whether a domain, function or quantity refers to the structure or respectively the fluid part of the system.

Let $\Omega \subset \mathbb{R}^2$ be a nonempty, closed and bounded Lipschitz domain and $[0, T] \subset \mathbb{R}$, $T > 0$ a time interval. In the following, the problem domain Ω consists of two time-dependent Lipschitz subdomains $\Omega_f(t)$ and $\Omega_s(t)$, with interface $\Gamma_I(t) := \overline{\Omega_f(t)} \cap \overline{\Omega_s(t)} = \partial\Omega_f(t) \cap \partial\Omega_s(t)$. For better readability we omit the time dependence of the domains and write Ω_f, Ω_s and Γ_I.

We further divide the remaining boundary of the fluid part, $\partial\Omega_f \setminus \Gamma_I$, into a Neumann part for the outflow condition Γ_N^{out}, and a Dirichlet part Γ_D^f. With Γ_D^{in} we denote the Dirichlet inflow boundary. For the solid part we only assume Dirichlet conditions on $\Gamma_D^s := \partial\Omega_s \setminus \Gamma_I$.

As a time independent reference domain we use the initial configuration of Ω, e.g., $\hat{\Omega}_f := \Omega_f(0)$ and $\hat{\Omega}_s := \Omega_s(0)$. Variables, functions, operators and quantities defined on the reference domain will always be indicated with the hat symbol $\hat{\cdot}$.

An example is the benchmark problem proposed by Turek and Hron [11], the setting is shown in Fig. 1. In the following we use the standard Hilbert spaces $L^2(G)$, $H^1(G)$ and the Sobolev spaces $W^{k,p}(G)$ for bounded Lipschitz domains $G \subset \mathbb{R}^2$,

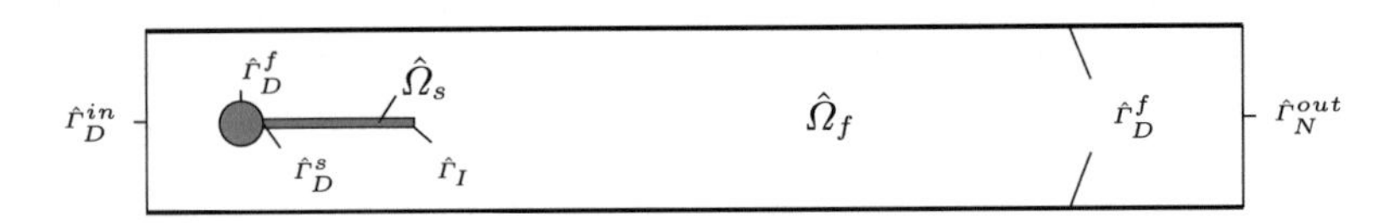

Fig. 1 Benchmark problem

for example $G = \Omega_s$, $G = \Omega_f$. We moreover define for $\Gamma_D \subset \partial G$

$$H^1_{0,D}(G) := \{f \in H^1(G) : f|_{\Gamma_D} = 0\}, \quad W^{k,p}_{0,D}(G) := \{f \in W^{k,p}(G) : f|_{\Gamma_D} = 0\},$$
$$C^\infty_{0,D}(\overline{G}) := \{f \in C^\infty(\overline{G}) : f|_{\Gamma_D} = 0\}.$$

Further, for the interface, $H^{1/2}_{00}(\Gamma_I)$ denotes the trace space of $H^1_{0,D}(\hat{\Omega}_f)$ on Γ_I.

We will sometimes omit the dimension d and write $L^2(G)$ for the product space $L^2(G)^d$, etc., and equip the spaces for $d > 1$ with the usual product norm.

For a Hilbert space H the inner product is denoted by $(\cdot,\cdot)_H$ and the associated norm by $\|\cdot\|_H = (\cdot,\cdot)_H^{1/2}$. For scalar product and norm in $L^2(G)$ we write shortly $(\cdot,\cdot)_G$ and $\|\cdot\|_G$. As we consider time dependent problems, we introduce for a space X the corresponding Bochner spaces $L^2((0,T);X)$ of square integrable functions from $(0,T)$ to X. If there exists a Hilbert space H, such that $X \hookrightarrow H \hookrightarrow X'$ with the dual space X' and continuous and dense imbeddings, we define

$$\mathcal{W}(X) := \{w \in L^2((0,T);X) \mid \partial_t w \in L^2((0,T);X')\},$$
$$\mathcal{W}_2(X) := \{w \in L^2((0,T);X) \mid \partial_{tt} w \in L^2((0,T);X')\}.$$

Moreover, we use the abbreviation $\mathcal{L}^2(X) = L^2((0,T);X)$.

3 Problem Setting

The hyperelastic structure is modeled in the Lagrangian approach given on the reference domain. With the momentum conservation equation, we obtain the following initial value problem for the deformation $\hat{u}$:

Find $\hat{u} : \hat{\Omega}_s \times [0,T] \to \mathbb{R}^2$ such that $\hat{u}(\cdot,0) = \hat{u}_0$, $\partial_t \hat{u}(\cdot,0) = \hat{u}_{0,t}$ in $\hat{\Omega}_s$ and

$$\rho_s \partial_t^2 \hat{u} - \widehat{\operatorname{div}}\,(\hat{J}\hat{\sigma}_s \hat{F}^{-T}) = \rho_s \hat{f}_s \quad \text{in} \quad \hat{\Omega}_s \times [0,T],$$
$$\hat{u} = 0 \quad \text{on} \quad \hat{\Gamma}^s_D \times [0,T],$$

where the density ρ_s is constant and $\hat{f}_s$ is a given right-hand side. $\hat{\mathcal{T}}$ describes the transformation $\hat{\mathcal{T}}(t,\hat{x}) := \hat{x} + \hat{u}(\hat{x},t)$, $\hat{F} := \widehat{\nabla}\hat{\mathcal{T}}$ its gradient and $\hat{J} := \det \hat{F}$. The stress tensor $\hat{\sigma}_s$ for a St. Venant-Kirchhoff material is given by

$$\hat{\sigma}_s := \hat{J}^{-1}\hat{F}(\lambda \operatorname{tr}(\hat{E})I + 2\mu\hat{E})\hat{F}^T,$$

with the Green-Lagrange strain tensor $\hat{E} = \frac{1}{2}(\hat{F}^T\hat{F} - I)$.

The equations for the fluid part are the two dimensional Navier-Stokes equations for incompressible flows. They yield the following initial value problem:

Find velocity $v : \Omega_f \times [0, T] \to \mathbb{R}^2$ and pressure $p : \Omega_f \times [0, T] \to \mathbb{R}$, such that $v(x, 0) = v_0(x)$ in Ω_f and:

$$\rho_f \partial_t v + \rho_f (v \cdot \nabla) v - \operatorname{div} \sigma_f = \rho_f f_f \quad \text{in} \quad \Omega_f \times [0, T],$$
$$\operatorname{div} v = 0 \quad \text{in} \quad \Omega_f \times [0, T],$$

with boundary conditions:

$$v = v^{in} \quad \text{on} \quad \Gamma_D^{in} \times [0, T],$$
$$v = 0 \quad \text{on} \quad \Gamma_D^{f} \times [0, T],$$
$$\rho_f \nu_f \nabla v \cdot n - pn = 0 \quad \text{on} \quad \Gamma_N^{out} \times [0, T],$$

where n denotes the unit normal outward vector. Again, the density ρ_f is constant, f_f is a given right-hand side and v^{in} is a constant inflow. The stress tensor σ_f is defined by

$$\sigma_f := -pI + \rho_f \nu_f (\nabla v + \nabla v^T),$$

with constant viscosity $\nu_f > 0$. At the Neumann outflow we prescribe the so-called *do-nothing* condition.

Note that the fluid problem is given on the current domain Ω_f in Eulerian coordinates. We will overcome this discrepancy regarding the Lagrangian coordinates of the structure with help of the *Arbitrary Lagrangian Eulerian* (ALE) formulation, see for example [2]. As ALE operator we extend the diffeomorphism of the deformation $\hat{\mathcal{T}}$ to $\hat{\Omega}_f$ and transform the equations onto the reference domain $\hat{\Omega}_f$:

$$\hat{J} \rho_f \partial_t \hat{v} + \hat{J} \rho_f (\hat{F}^{-1}(\hat{v} - \partial_t \hat{\mathcal{T}}) \cdot \hat{\nabla}) \hat{v} - \widehat{\operatorname{div}}\, (\hat{J} \hat{\sigma}_f \hat{F}^{-T}) = \hat{J} \rho_f \hat{f}_f \quad \text{in } \hat{\Omega}_f \times [0, T],$$
$$\widehat{\operatorname{div}}\, (\hat{J} \hat{F}^{-1} \hat{v}) = 0 \quad \text{in } \hat{\Omega}_f \times [0, T].$$

The transformed Cauchy stress tensor is given by

$$\hat{\sigma}_f = -\hat{p}\hat{I} + \rho_f \nu_f (\hat{\nabla} \hat{v} \hat{F}^{-1} + \hat{F}^{-T} \hat{\nabla} \hat{v}^T).$$

For the boundary we assume that $\partial \Omega_f \setminus \Gamma_I = \partial \hat{\Omega}_f \setminus \hat{\Gamma}_I$, which is reasonable, see for example the benchmark problem in Fig. 1.

The interaction conditions at the fluid-solid interface $\hat{\Gamma}_I$ are

$$\partial_t \hat{u} = \hat{v} \quad \text{on} \quad \hat{\Gamma}_I \times [0, T], \tag{1}$$
$$\hat{J} \hat{\sigma}_f \hat{F}^{-T} \hat{n}_f = \hat{J} \hat{\sigma}_s \hat{F}^{-T} \hat{n}_s \quad \text{on} \quad \hat{\Gamma}_I \times [0, T], \tag{2}$$

where $\hat{n}_f$ and $\hat{n}_s$ are the unit normal outward vectors. The first equation is called the continuity of velocities, as the time derivative of a displacement is again a velocity. The second condition is the balance of forces across the interface.

Now, we can state the variational formulation of the problem completely on the reference domain.

Problem 1 Let $\hat{v}_D : \Omega_f \times [0, T] \to \mathbb{R}^2$ be an extension of the boundary data, i.e., $\hat{v}|_{\hat{\Gamma}_D^f} = 0$, $\hat{v}|_{\hat{\Gamma}_D^{in}} = \hat{v}^{in}$. Find $(\hat{u}, \hat{v}, \hat{p}, \hat{e}) \in \hat{Y}_D := (0, \hat{v}_D, 0, 0) + \hat{Y}$, where

$$\hat{Y} \subset \mathcal{W}_2(W^{1,4}_{0,D}(\hat{\Omega}_s)^2) \times \mathcal{W}(H^1_{0,D}(\hat{\Omega}_f)^2) \times \mathcal{L}^2(L^2(\hat{\Omega}_f)) \times \mathcal{L}^2((H^{1/2}_{00}(\hat{\Gamma}_I)^2)')$$

is a suitable Banach space, such that for almost all $t \in [0, T]$:

$$\begin{aligned}
(\rho_s \partial_t^2 \hat{u}, \hat{\phi}_s)_{\hat{\Omega}_s} + (\hat{F}(\lambda \operatorname{tr}(\hat{E}) I + 2\mu \hat{E}), \hat{\nabla}\hat{\phi}_s)_{\hat{\Omega}_s} + (\hat{e}, \hat{\phi}_s)_{\hat{\Gamma}_I}& \\
-(\rho_s \hat{f}_s, \hat{\phi}_s)_{\hat{\Omega}_s} &= 0, \\
(\rho_f \hat{J} \partial_t \hat{v}, \hat{\phi}_f)_{\hat{\Omega}_f} + (\rho_f \hat{J}(\hat{F}^{-1}(\hat{v} - \partial_t \hat{\mathcal{T}}) \cdot \hat{\nabla})\hat{v}, \hat{\phi}_f)_{\hat{\Omega}_f} + (\hat{J}\hat{\sigma}_f \hat{F}^{-T}, \hat{\nabla}\hat{\phi}_f)_{\hat{\Omega}_f}& \\
-(\hat{J}\rho_f \nu_f \hat{F}^{-T} \hat{\nabla}\hat{v}^T \hat{F}^{-T} \hat{n}_f, \hat{\phi}_f)_{\hat{\Gamma}_N^{out}} - (\hat{e}, \hat{\phi}_f)_{\hat{\Gamma}_I} - (\rho_f \hat{J} \hat{f}_f, \hat{\phi}_f)_{\hat{\Omega}_f} &= 0, \\
(\widehat{\operatorname{div}}\,(\hat{J}\hat{F}^{-1}\hat{v}), \hat{\phi}_p)_{\hat{\Omega}_f} &= 0, \\
(\partial_t \hat{u} - \hat{v}, \hat{\phi}_e)_{\hat{\Gamma}_I} &= 0, \\
\hat{u}(0) - \hat{u}_0 &= 0, \\
\hat{v}(0) - \hat{v}_0 &= 0,
\end{aligned}$$

for all $(\hat{\phi}_s, \hat{\phi}_f, \hat{\phi}_p, \hat{\phi}_e) \in C^\infty_{0,D}(\overline{\hat{\Omega}_s})^2 \times C^\infty_{0,D}(\overline{\hat{\Omega}_f})^2 \times C^\infty_0(\overline{\hat{\Omega}_f}) \times C^\infty_0(\overline{\hat{\Gamma}_I})^2$. For the system we write shortly

$$\hat{\mathcal{S}}(\hat{y}) = 0,$$

for $\hat{y} := (\hat{u}, \hat{v}, \hat{p}, \hat{e}) \in (0, \hat{v}_D, 0, 0) + \hat{Y}$.

The existence and uniqueness for solutions of Problem 1 is very involved and beyond the scope of this paper. Recently, under suitable assumptions on the geometry and on the data regularity the local existence of a unique solution in the product of suitable fractional Sobolev spaces was shown in [8] for the case of linear elasticity instead of the St. Venant-Kirchhoff model.

We consider now optimal control problems of the form

$$\begin{aligned}
\min_{\hat{y} \in \hat{Y}_D, \hat{c} \in \hat{C}_{ad}} \quad & \hat{\mathcal{J}}(\hat{y}(\hat{c}), \hat{c}) \\
\text{s.t.} \quad & \hat{\mathcal{S}}(\hat{y}, \hat{c}) = 0,
\end{aligned}$$

with the additional control $\hat{c}$, from a suitable control space $\hat{C}$, $\hat{C}_{ad} \subset \hat{C}$ closed and convex, and a cost functional $\hat{\mathcal{J}} : \hat{Y}_D \times \hat{C} \to \mathbb{R}$ depending on state and control.

4 Discretization

In this section we give a short overview of the underlying monolithic discretization, which is similar to the solver proposed in [6]. For the time discretization we use a Crank-Nicolson scheme and therefore partition the time interval into N intervals with length $\Delta t > 0$, i.e., $N\Delta t = T$. We use a staggered time grid in the sense that $\hat{v}_k^h$ is the approximation at $k\Delta t$, and $\hat{u}_k^k$ is the displacement approximation at $(k + \frac{1}{2})\Delta t$.

The space is discretized with a finite element approach. Therefore we choose a regular triangulation for a polygonal approximation $\hat{\Omega}^h$ of $\hat{\Omega}$ leading to subdomains $\hat{\Omega}_f^h$ and $\hat{\Omega}_s^h$ for the fluid and the structure with corresponding boundaries. On these meshes, we define finite element spaces. For the fluid we consider the well-known Taylor-Hood elements. Hence, for the velocity we choose continuous P2-elements, which induce a space $V^h(\hat{\Omega}_f^h) \subset H_{0,D}^1(\hat{\Omega}_f^h)$. The pressure is approximated with continuous P1-elements $L^h(\hat{\Omega}_f^h) \subset L^2(\hat{\Omega}_f^h)$. As ansatz functions for the deformation, we also choose continuous P2-elements $V^h(\hat{\Omega}_s) \subset W_{0,D}^{1,4}(\hat{\Omega}_s^h)$, and moreover the interface variable e is approximated by continuous piecewise quadratic elements on $\hat{\Gamma}_I$, i.e., $L^h(\hat{\Gamma}_I) \subset L^2(\hat{\Gamma}_I)$.

Moreover, let $\hat{v}_{k,D}^h$ be an approximation of the boundary data $\hat{v}_D(k\Delta t, \cdot)$ in the space of continuous P2-elements. We use $\hat{Y}_{D,k}^h := (0, \hat{v}_{D,k}^h, 0, 0) + \hat{Y}^h$, where

$$\hat{Y}^h := V^h(\hat{\Omega}_s) \times V^h(\hat{\Omega}_f) \times L^h(\hat{\Omega}_f) \times L^h(\hat{\Gamma}_I)$$

as ansatz space and $\hat{Y}^h$ as test space for time step k.

The discrete operator $\hat{\mathcal{T}}^h$ is a continuation of the interface displacement into the fluid domain. We choose a piecewise linear continuation on the pressure grid with a given linear extension operator A_{ext}, i.e.,

$$\hat{\mathcal{T}}^h(\hat{u}_k^h, \hat{u}_{k-2}^h) := A_{ext} \frac{\hat{u}_k^h|_{\hat{\Gamma}_I} + \hat{u}_{k-2}^h|_{\hat{\Gamma}_I}}{2},$$

and define the mesh velocity

$$\partial_t^h \hat{\mathcal{T}}^h(\hat{u}_k^h, \hat{u}_{k-2}^h) := A_{ext} \frac{\hat{u}_k^h|_{\hat{\Gamma}_I} - \hat{u}_{k-2}^h|_{\hat{\Gamma}_I}}{2\Delta t}.$$

Finally, let $\hat{u}_0^h, \frac{\hat{u}_0^h - \hat{u}_{-1}^h}{\Delta t} \in V^h(\hat{\Omega}_s)$ and $\hat{v}_0^h \in \hat{v}_{D,0}^h + V^h(\hat{\Omega}_f)$ be approximations of $\hat{u}_0$, $\hat{u}_{0,t}$ and $\hat{v}_0$, respectively.

Now we can state the fully discretized problem.

Problem 2 For $k = 1, 2, \dots N$ find $(\hat{u}_k^h, \hat{v}_k^h, \hat{p}_k^h, \hat{e}_k^h) \in \hat{Y}_{D,k}^h$ such that:

$$\frac{\rho_s}{\Delta t^2}(\hat{u}_k^h - 2\hat{u}_{k-1}^h + \hat{u}_{k-2}^h, \hat{\phi}_s^h)_{\hat{\Omega}_s^h} + (\hat{F}_m^h(\lambda\,\mathrm{tr}(\hat{E}_m^h)I + 2\mu\hat{E}_m^h), \hat{\nabla}\phi_s^h)_{\hat{\Omega}_s^h}$$

$$+(\hat{e}_k^h, \hat{\phi}_s^h)_{\hat{\Gamma}_I^h} - (\rho_s \hat{f}_{s,k}^h, \hat{\phi}_s^h)_{\hat{\Omega}_s^h} = 0,$$

$$\rho_f \hat{J}_m(\frac{1}{\Delta t}(\hat{v}_k^h - \hat{v}_{k-1}^h) + (\hat{F}_m^{-1}(\hat{v}_m^h - \partial_t^h \hat{\mathcal{T}}^h(\hat{u}_k^h, \hat{u}_{k-2}^h)) \cdot \hat{\nabla})\hat{v}_m^h, \hat{\phi}_f^h)_{\hat{\Omega}_f^h} - (\hat{e}_k^h, \hat{\phi}_f^h)_{\hat{\Gamma}_I^h}$$

$$-\hat{J}_m((\hat{p}_k^h I - \rho_f \nu_f(\hat{\nabla}\hat{v}_m^h \hat{F}_m^{-1} + \hat{F}_m^{-T}(\hat{\nabla}\hat{v}_m^h)^T))\hat{F}_m^{-T}, \hat{\nabla}\hat{\phi}_f^h)_{\hat{\Omega}_f^h} - (\rho_f \hat{J}_m \hat{f}_{f,k}^h, \hat{\phi}_f^h)_{\hat{\Omega}_f^h}$$

$$-(\rho_f \nu_f \hat{J}_m \hat{F}_m^{-T}(\hat{\nabla}\hat{v}^h)^T \hat{n}_f^h, \hat{\phi}_f^h)_{\hat{\Gamma}_N^{h,out}} + (\frac{1}{2}\hat{v}_m^h \widehat{\mathrm{div}}(\hat{J}\hat{F}^{-1}\hat{v}_m^h), \hat{\phi}_f^h)_{\hat{\Omega}_f} = 0,$$

$$(\widehat{\mathrm{div}}\,(\hat{J}_m \hat{F}_m^{-1}\hat{v}_k^h), \hat{\phi}_p^h)_{\hat{\Omega}_f^h} = 0,$$

$$(\frac{1}{\Delta t}(\hat{u}_k^h - \hat{u}_{k-1}^h) - \hat{v}_k^h, \phi_e^h)_{\hat{\Gamma}_I} = 0,$$

for all $(\hat{\phi}_s^h, \hat{\phi}_f^h, \hat{\phi}_p^h, \hat{\phi}_e^h) \in \hat{Y}^h$. Here, $\hat{u}_m^h := \frac{1}{2}(\hat{u}_k^h + \hat{u}_{k-2}^h)$, $\hat{v}_m^h := \frac{1}{2}(\hat{v}_k^h + \hat{v}_{k-1}^h)$ and $\hat{F}_m$ is shorthand for $\hat{F}(\hat{u}_m^h)$, $\hat{J}_m$ for $\hat{J}(\hat{u}_m^h)$ and $\hat{E}_m$ for $\hat{E}(\hat{u}_m^h)$. We further have added a stabilization term to the fluid equation.

For this fully discretized problem we write shortly

$$\mathcal{S}^h(\hat{y}^h) = 0,$$

with $\hat{y}^h = (\hat{u}_k^h, \hat{v}_k^h, \hat{p}_k^h, \hat{e}_k^h)_{1 \le k \le N}$.

We identify a finite element function $\hat{y}^h$ with its coefficient vector $\mathbf{y} \in \mathbb{R}^{n_y}$ from the unique representation

$$\hat{y}^h = \hat{y}_D^h + \sum_{j=1}^{n_y} \mathbf{y}_j \hat{\psi}_j,$$

where $\hat{\psi}_j$ are the finite element basis functions, n_y is the number of degrees of freedom and $\hat{y}_D^h$ is a nonhomogeneous part. In matrix form, after rearranging the equations we have

$$
\begin{pmatrix}
\frac{\rho_s}{\Delta t^2} M_s + \frac{1}{2} S_s & & & I_s^T N \\
& \frac{\rho_f}{\Delta t} M_f + \frac{1}{2} S_f & -D^T & -I_f^T N \\
& -D & & \\
\frac{1}{\Delta t} N I_s & -N I_f & &
\end{pmatrix}
\begin{pmatrix} \mathbf{u}_k \\ \mathbf{v}_k \\ \mathbf{p}_k \\ \mathbf{e}_k \end{pmatrix} =
\begin{pmatrix}
\rho_s M_s \mathbf{f}_s - \frac{1}{2} S_s \mathbf{u}_{k-1} + \frac{\rho_s}{\Delta t^2} M_s (2\mathbf{u}_{k-1} - \mathbf{u}_{k-2}) \\
\rho_f M_f \mathbf{f}_f + (\frac{\rho_f}{\Delta t} M_f - \frac{1}{2} S_f) \mathbf{v}_{k-1} \\
0 \\
\frac{1}{\Delta t} N I_s \mathbf{u}_{k-1}
\end{pmatrix} + \mathcal{F}_k, \tag{3}
$$

where $\mathcal{F}_k$ results from the inhomogeneous Dirichlet data $\hat{v}_{D,k}^h$, M_s, M_f denote the mass matrices of the solid and the fluid domain respectively, N denotes the mass matrix on the interface, and I_s, I_f are the restrictions of functions defined on the solid or the fluid domain on Γ_I. S_s denotes the system matrix for the solid, that means

$$
\boldsymbol{\phi}_s^T S_s \frac{(\mathbf{u}_k + \mathbf{u}_{k-2})}{2} = (\hat{F}_m^h (\lambda \operatorname{tr}(\hat{E}_m^h) I + 2\mu \hat{E}_m^h), \hat{\nabla} \phi_s^h)_{\hat{\Omega}_s^h}
$$

and S_f the system matrix for the fluid:

$$
\begin{aligned}
\boldsymbol{\phi}_f^T S_f \frac{(\mathbf{v}_k + \mathbf{v}_{k-1})}{2} &= \hat{J}_m ((\hat{F}_m^{-1} (\hat{v}_m^h - \partial \hat{\mathcal{T}}_m^h) \cdot \hat{\nabla}) \hat{v}_m^h, \hat{\phi}_f^h)_{\hat{\Omega}_f^h} \\
&+ \hat{J}_m (\rho_f \nu_f (\hat{\nabla} \hat{v}_m^h \hat{F}_m^{-1} - \hat{F}_m^{-T} (\hat{\nabla} \hat{v}_m^h)^T)) \hat{F}_m^{-T}, \hat{\nabla} \hat{\phi}_f^h)_{\hat{\Omega}_f^h} \\
&- (\rho_f \nu_f \hat{J}_m \hat{F}_m^{-T} (\hat{\nabla} \hat{v}^h)^T \hat{n}_f^h, \hat{\phi}_f^h)_{\hat{\Gamma}_N^{h,out}} + (\frac{1}{2} \hat{v}_m^h \widehat{\operatorname{div}} (\hat{J} \hat{F}^{-1} \hat{v}_m^h), \hat{\phi}_f^h)_{\hat{\Omega}_f}
\end{aligned}
$$

further D is the divergence matrix defined via

$$
\boldsymbol{\phi}_p^T D \mathbf{v}_k = (\widehat{\operatorname{div}} (\hat{J}_m \hat{F}_m^{-1} \hat{v}_k^h), \hat{\phi}_p^h)_{\hat{\Omega}_f^h}.
$$

The discretized optimization problem is given by

$$
\begin{aligned}
\min_{\hat{y}^h \in \hat{Y}_D^h, \hat{c}^h \in \hat{C}_{ad}^h} \quad & \hat{\mathcal{J}}(\hat{y}^h(\hat{c}^h), \hat{c}^h) \\
\text{s.t.} \quad & \mathcal{S}^h(\hat{y}^h, \hat{c}^h) = 0
\end{aligned}
$$

with a cost functional $\hat{\mathcal{J}} : \hat{Y}_D^h \times \hat{C}^h \to \mathbb{R}$ depending on the state and a control $\hat{c}^h$ from a suitable control space $\hat{C}^h$, with $\hat{C}_{ad} \subset \hat{C}^h$ closed and convex.

5 Model Order Reduction via POD

The main idea is to approximate the finite element spaces with reduced spaces to lower the dimension of the system and reduce the computational time of the optimization.

For the model reduction we use a proper orthogonal decomposition (POD) technique, see for example [7]. It has already been successfully used for a variety of optimal control problems, e.g., [4, 12]. We will now shortly revise the discrete variant, before we introduce the model for FSI.

For this purpose, let H be a real Hilbert space. Snapshots are elements $y_1, \dots, y_n \in H$ for $n \in \mathbb{N}$, with at least one non-zero element. We call the span of the snapshots $V := \text{span}\{y_j | 1 \leq j \leq n\}$ the snapshot subspace. It is a linear subspace of H with dimension $d \leq n$.

Now the idea is to choose an orthonormal basis $\{\psi_i\}_{i=1,\dots,l} \in H$, which solves the following minimization problem

$$\begin{aligned} &\min_{\{\psi_i\}_{i=1}^l} \sum_{j=1}^n \|y_i - \sum_{i=1}^l (y_j, \psi_i)_H \psi_i\|_H^2 \\ &\text{s.t. } \{\psi_i\}_{i=1}^l \subset H \text{ and } (\psi_i, \psi_j)_H = \delta_{ij},\ 1 \leq i, j \leq l, \end{aligned} \tag{4}$$

where the dimension l of the POD space is given and δ_{ij} denotes the Kronecker delta. We note that the second term in the norm is the Fourier sum of the snapshots in the approximated space. The constraint guarantees the orthogonality of the basis vectors in H. The space $V^r := \text{span}\{\psi_1, \dots, \psi_l\} \subset H$ is called the POD subspace.

Let us now w.l.o.g. assume that the snapshots are linearly independent, i.e., $d = n$. If they are not, we can reduce the snapshot set until it is linear independent without losing information. Then, the minimization problem (4) is equivalent to the $n \times n$ eigenvalue problem of the correlation matrix $K = (K_{ij}) \in \mathbb{R}^{n \times n}$ with $K_{ij} = (y_i, y_j)_H$. With $\tilde{a}_k \in \mathbb{R}^n$ we can rewrite it as

$$K\tilde{a}_k = \lambda_k \tilde{a}_k,$$

where K is positive definite. We assume the eigenvalues to be in descending order. The basis vectors can be computed via

$$\psi_k := \frac{1}{\sqrt{\lambda_k}} \sum_{i=1}^n (\tilde{a}_k)_i y_i,$$

for $k = 1, \dots, l$.

6 Proper Orthogonal Decomposition for FSI

Since we have four different variables and equations, we will compute four different POD basis.

6.1 *Fluid: Velocity and Pressure*

To treat the Dirichlet boundary conditions, we follow [3] and calculate a mean flow field of the time snapshots of the velocity $\hat{v}_k^h \in \hat{v}_{D,k}^h + V^h(\hat{\Omega}_f^h)$, $1 \leq k \leq n$,

$$\bar{v}^h = \frac{1}{n}\sum_{i=1}^{n} \hat{v}_i^h.$$

We then apply the POD method to the following modified set of snapshots

$$V_v = \text{span}\{\hat{y}_1^h, \ldots, \hat{y}_n^h\}, \quad \text{where} \quad \hat{y}_i^h = \hat{v}_i^h - \bar{v}^h \quad \text{for all} \quad i = 1, \ldots, n.$$

If we have a constant inflow, these modified snapshots fulfill homogeneous Dirichlet boundary conditions. To construct the reduced basis, we choose $H = H_{0,D}^1(\hat{\Omega}_f^h)$ with inner product $(\cdot,\cdot)_{H_{0,D}^1(\hat{\Omega}_f)}$.

This yields an orthogonal basis $\Psi^v := \{\hat{\psi}_1^v, \ldots, \hat{\psi}_{l_v}^v\}$ in $H_{0,D}^1(\hat{\Omega}_f)$. We can then describe the fluid velocity via

$$\hat{v}^r(\hat{x}, t) := \bar{v}^h(\hat{x}) + \sum_{j=1}^{l_v} \beta_j^v(t)\hat{\psi}_j^v(\hat{x}) \quad \text{for } \hat{x} \in \hat{\Omega}_f, \quad t \in [0, T]$$

with the time-independent orthogonal basis functions $\hat{\psi}_j^v$ and corresponding time coefficients β_j^v, for $j = 1, \ldots, l_v$.

We define the POD subspace for the velocity by $V_v^r := \text{span}\{\Psi^v\}$ and its dimension is $l_v \leq n \ll n_v$.

Now let us consider the pressure $\hat{p}^h$. We calculate finite element time snapshots $\hat{p}_1^h, \ldots, \hat{p}_n^h \in L^h(\hat{\Omega}_f)$ and get the snapshot subspace

$$V_p := \text{span}\{\hat{p}_1^h, \ldots, \hat{p}_n^h\}.$$

For the POD basis calculation we choose $H = L^2(\hat{\Omega}_f)$, with inner product $(\cdot,\cdot)_{\hat{\Omega}_f}$. Then we get a POD basis $\Psi^p := \{\hat{\psi}_1^p, \ldots, \hat{\psi}_{l_p}^p\}$ of the POD subspace V_p^r for the pressure with dimension $l_p \leq n \ll n_p$. The corresponding temporal dependent variables are denoted by β_j^p so that all in all we get a reduced variable

$$\hat{p}^r(\hat{x}, t) := \sum_{j=1}^{l_p} \beta_j^p(t)\hat{\psi}_j^p(\hat{x}) \quad \text{for } \hat{x} \in \hat{\Omega}_f, \quad t \in [0, T].$$

6.2 Solid: Deformation

The snapshot space $V_u := \text{span}\{\hat{u}_1^h, \dots, \hat{u}_n^h\}$ of the deformation is a subspace of $W_{0,D}^{1,4}(\hat{\Omega}_s) \subset H_{0,D}^1(\hat{\Omega}_s)$, so we choose $H = H_{0,D}^1(\hat{\Omega}_s)$. The resulting POD basis of rank $l_u \leq n \ll n_u$ is further denoted by $\Psi^u := \{\hat{\psi}_1^u, \dots, \hat{\psi}_{l_u}^u\}$, and its span by V_u^r. The corresponding time dependent variables are consequently denoted by β^u such that we obtain the reduced variable $\hat{u}^r$

$$\hat{u}^r(\hat{x}, t) := \sum_{j=1}^{l_u} \beta_j^u(t)\hat{\psi}_j^u(\hat{x}) \quad \text{for } \hat{x} \in \hat{\Omega}_s, \quad t \in [0, T].$$

6.3 Interface Variable

The derivation of the reduced variable e^r is analogously. The subspace $V_e := \text{span}\{e_1, \dots, e_n\}$ spanned by the snapshots of the interface variable is a subspace of $L^h(\hat{\Gamma}_I^h) \subset L^2(\hat{\Gamma}_I^h)$. Since the results in [8] indicate that for sufficiently regular data the normal stresses lie in $L^2(\hat{\Gamma}_I^h)$ for a.a. times, we choose $H = L^2(\hat{\Gamma}_I^h)$ and obtain a basis of the POD subspace V_e^r with rank $l_e \leq n \ll n_e$, which we denote by Ψ^e. Together with time dependent coefficients β_j^e we arrive at

$$\hat{e}^r(\hat{x}, t) := \sum_{j=1}^{l_e} \beta_j^e(t)\hat{\psi}_j^e(\hat{x}) \quad \text{for } \hat{x} \in \hat{\Gamma}_I, \quad t \in [0, T].$$

6.4 Dimension of the POD Subspaces

Let λ_i, $i = 1, \dots, n$, be the eigenvalues of K and $\lambda_1 \geq \dots \geq \lambda_n$. We choose the smallest l, such that

$$\frac{\sum_{i=1}^{l} \lambda_i}{\sum_{i=1}^{n} \lambda_i} \geq \delta,$$

for given $\delta \in (0, 1]$, where we typically choose δ close to one. We call δ the preserved energy of the system because it gives us a measure on how much information of the snapshot set is preserved in the reduced order model.

6.5 Reduced Problem

Now we can state the reduced problem:

Problem 3 For $k = 1, \dots, N$ find deformation $\hat{u}_k^r \in V_u^r$, velocity $\hat{v}_k^r \in \bar{v} + V_v^r$, pressure $\hat{p}_k^r \in V_p^r$, and interface variable $\hat{e}_k^r \in V_e^r$, such that

$$\frac{\rho_s}{\Delta t^2}(\hat{u}_k^r - 2\hat{u}_{k-1}^r + \hat{u}_{k-2}^r, \hat{\phi}_s^r)_{\hat{\Omega}_s^h} + (\hat{F}_m^r(\lambda \operatorname{tr}(\hat{E}_m^r)I + 2\mu \hat{E}_m^r), \hat{\nabla}\phi_s^r)_{\hat{\Omega}_s^h}$$
$$+(\hat{e}_k^r, \hat{\phi}_s^r)_{\hat{\Gamma}_I^h} - (\rho_s \hat{f}_s^{r,k}, \hat{\phi}_s^r)_{\hat{\Omega}_s^h} = 0,$$

$$\rho_f \hat{J}_m^r(\frac{1}{\Delta t}(\hat{v}_k^r - \hat{v}_{k-1}^r) + ((\hat{F}_m^r)^{-1}(\hat{v}_m^r - \partial_t^h \hat{\mathcal{T}}^h(\hat{u}_k^r, \hat{u}_{k-2}^r)) \cdot \hat{\nabla})\hat{v}_m^r - \hat{f}_{f,k}^r, \hat{\phi}_f^r)_{\hat{\Omega}_f^h} -$$
$$(\hat{e}_k^r, \hat{\phi}_s^r)_{\hat{\Gamma}_I^h} - \hat{J}_m^r(\hat{p}_k^r I - \rho_f \nu_f(\hat{\nabla}\hat{v}_m^r(\hat{F}_m^r)^{-1} + (\hat{F}_m^r)^{-T}(\hat{\nabla}\hat{v}_m^r)^T)(\hat{F}_m^r)^{-T}, \hat{\nabla}\hat{\phi}_f^r)_{\hat{\Omega}_f^h}$$
$$-(\rho_f \nu_f \hat{J}_m^r(\hat{F}_m^r)^{-T}(\hat{\nabla}\hat{v}^r)^T \hat{n}_f^h, \hat{\phi}_f^r)_{\hat{\Gamma}_N^{h,out}} + (\frac{1}{2}\hat{v}_m^r \widehat{\operatorname{div}}(\hat{J}_m^r(\hat{F}_m^r)^{-1}\hat{v}_m^r), \hat{\phi}_f^r)_{\hat{\Omega}_f^h} = 0,$$
$$(\widehat{\operatorname{div}}\,(\hat{J}_m^r(\hat{F}_m^r)^{-1}\hat{v}_k^r), \hat{\phi}_p^r)_{\hat{\Omega}_f^h} = 0,$$
$$(\frac{1}{\Delta t}(\hat{u}_k^r - \hat{u}_{k-1}^r) - \hat{v}_m^r, \psi^r)_{\hat{\Gamma}_I^h} = 0,$$

for all $(\hat{\phi}_s^r, \phi_f^r, \phi_p^r, \hat{\psi}^r) \in \hat{Y}^r := V_u^r \times V_v^r \times V_p^r \times V_e^r$. Here, $\hat{u}_0^r$, $\hat{u}_{-1}^r$, and $\hat{v}_0^h$ are the projected initial data, $\hat{u}_m^r := \frac{1}{2}(\hat{u}_k^r + \hat{u}_{k-2}^r)$, $\hat{v}_m^r := \frac{1}{2}(\hat{v}_k^r + \hat{v}_{k-1}^r)$. $\hat{F}_m^r$ is a shorthand notation for $\hat{F}(\hat{u}_m^r)$, $\hat{J}_m^r$ for $\hat{J}(\hat{u}_m^r)$, and $\hat{E}_m$ for $\hat{E}(u_m^r)$.

We further write analogously to before

$$\mathcal{S}^r(\hat{y}^r) = 0,$$

with $\hat{y}^r = (\hat{u}_k^r, \hat{v}_k^r, \hat{p}_k^r, \hat{e}_k^r)_{1 \le k \le N}$.

The matrix form is with the definitions from (3), $[\Psi^y] = [\hat{\psi}_1^y, \dots, \hat{\psi}_{l_y}^y]$, $y \in \{u, v, p, e\}$, and

$$\tilde{M}_s := [\Psi^u]^T M_s [\Psi^u], \quad \tilde{M}_f := [\Psi^v]^T M_f [\Psi^v],$$
$$\tilde{N}_s := [\Psi^u]^T I_s^T N [\Psi^e], \quad \tilde{N}_f := [\Psi^v]^T I_f^T N [\Psi^e],$$
$$\tilde{D} := [\Psi^f]^T D [\Psi^p]$$

given by

$$\begin{pmatrix} \frac{\rho_s}{\Delta t^2}\tilde{M}_s + [\Psi^s]^T \frac{1}{2} S_s [\Psi^s] & & & \tilde{N}_s \\ & \frac{\rho_f}{\Delta t}\tilde{M}_f + [\Psi^v]^T \frac{1}{2} S_f [\Psi^v] & -\tilde{D}^T & -\tilde{N}_f \\ & -\tilde{D} & & \\ \frac{1}{\Delta t}\tilde{N}_s^T & -\tilde{N}_f^T & & \end{pmatrix} \begin{pmatrix} \boldsymbol{\beta}_k^u \\ \boldsymbol{\beta}_k^v \\ \boldsymbol{\beta}_k^p \\ \boldsymbol{\beta}_k^e \end{pmatrix} =$$

$$\begin{pmatrix} \rho_s [\Psi^u]^T M_s \mathbf{f}_s - \frac{1}{2}[\Psi^u]^T S_s \mathbf{u}_{k-1}^r + \frac{\rho_s}{\Delta t^2}[\Psi^u]^T M_s (2\mathbf{u}_{k-1}^r - \mathbf{u}_{k-2}^r) \\ \rho_f [\Psi^v]^T M_f \mathbf{f}_f + [\Psi^u]^T (\frac{\rho_f}{\Delta t} M_f - \frac{1}{2} S_f)\mathbf{v}_{k-1} \\ 0 \\ \frac{1}{\Delta t}[\Psi^e]^T N I_s \mathbf{u}_{k-1} \end{pmatrix} - r(\bar{v}), \tag{5}$$

where

$$r(\bar{v}) = \begin{pmatrix} 0 \\ [\Psi^v]^T (\frac{\rho_f}{\Delta t}\tilde{M}_f + \frac{1}{2} S_f)\bar{v} \\ -[\Psi^p]^T D\bar{v} \\ [\Psi^e]^T N I_f \bar{v} \end{pmatrix}$$

is the contribution of the mean flow field of the snapshots. Equivalently to before we define the optimization problem:

$$\min_{\hat{y}^r \in \hat{Y}^r, \hat{c}^h \in \hat{C}^h_{ad}} \hat{\mathcal{J}}(\hat{y}^r(\hat{c}^h), \hat{c}^h)$$

$$\text{s.t. } \mathcal{S}^r(\hat{y}^r, \hat{c}^h) = 0$$

for the reduced order model.

7 Enrichment by Supremizers

If we divide the solid equation by Δt, the reduced system (5) induces a saddle point problem for every k. For existence and stability of solutions, we want to make sure that the discrete *inf-sup* condition:

$$\inf_{\phi^h \in H_2} \sup_{y^h \in H_1} \frac{b(y^h, \phi^h)}{\|y^h\|_{H_1} \|\phi^h\|_{H_2}} \geq \gamma > 0,$$

for the corresponding bilinear form b and a given constant γ, holds true for the derived reduced order model.

This does not need to be the case for the standard choices of the POD basis. Thus, we follow the proposal of [10] and enrich the POD basis of velocity and deformation with information of pressure and interface variables. The basic idea is to enrich the POD basis of velocity and deformation:

$$\Psi_1 := \{(\hat{\psi}_i^u, 0), (0, \hat{\psi}_j^v) : 1 \le i \le l_u, 1 \le j \le l_v\}$$

of $H_1^r := V_u^r \times V_v^r$, used to approximate $H_1 := V^h(\hat{\Omega}_s^h) \times V^h(\hat{\Omega}_f^h)$, with information from the POD basis

$$\Psi_2 := \{(\hat{\psi}_k^p, 0), (0, \hat{\psi}_l^e) : 1 \le k \le l_p, 1 \le l \le l_e\}$$

of $H_2^r := V_p^r \times V_e^r$ used to approximate $H_2 := L^h(\hat{\Omega}_f^h) \times L^h(\hat{\Gamma}_I^h)$. We set

$$H_1^{r,s} := \text{span}\{\Psi_1, \mathcal{O}\Psi_2\},$$

where $\mathcal{O}\Psi_2$ is the short notation for $\{\mathcal{O}\psi_i, \forall \psi_i \in \Psi_2\}$. The operator $\mathcal{O} : H_2 \to H_1$ is defined by the bilinear form b via the scalar product in H_1:

$$b(\hat{y}^h, \hat{\phi}^h) = (\mathcal{O}\hat{\phi}^h, \hat{y}^h)_{H_1} \quad \text{for } \hat{y}^h = (\hat{u}^h, \hat{v}^h) \in H_1, \; \hat{\phi}^h = (\hat{\phi}_p^h, \hat{\phi}_e^h) \in H_2.$$

Now, we see that if the finite element approach satisfies the *inf-sup* condition, so does the reduced order model:

By construction $\mathcal{O}\psi^r \in H_1^r$ for all $\psi^r \in H_2^r$ and thus

$$\begin{aligned}
\gamma^r &:= \inf_{\psi^r \in H_2^r} \sup_{v^r \in H_1^{r,s}} \frac{b(v^r, \psi^r)}{\|v^r\|_{H_1} \|\psi^r\|_{H_2}} = \inf_{\psi^r \in H_2^r} \sup_{v^r \in H_1^{r,s}} \frac{(v^r, \mathcal{O}\psi^r)_{H_1}}{\|v^r\|_{H_1} \|\psi^r\|_{H_2}} \\
&= \inf_{\psi^r \in H_2^r} \frac{(\mathcal{O}\psi^r, \mathcal{O}\psi^r)_{H_1}}{\|\mathcal{O}\psi^r\|_{H_1} \|\psi^r\|_{H_2}} \ge \inf_{\psi \in H_2} \frac{(\mathcal{O}\psi, \mathcal{O}\psi)_{H_1}}{\|\mathcal{O}\psi\|_{H_1} \|\psi\|_{H_2}} \\
&= \inf_{\psi \in H_2} \sup_{v \in H_1} \frac{(v, \mathcal{O}\psi)_{H_1}}{\|v\|_{H_1} \|\psi\|_{H_2}} = \gamma > 0,
\end{aligned}$$

where we have used the Cauchy-Schwarz inequality in the second line.

We have

$$b(y^h, \phi^h) = \frac{1}{\Delta t} (\hat{u}_k^h, \hat{\phi}_e^h)_{\hat{\Gamma}_I^h} - (\hat{v}_k^h, \hat{\phi}_e)_{\hat{\Gamma}_I^h} - (\widehat{\text{div}}(\hat{J}\hat{F}^{-1}\hat{v}_k^h), \hat{\phi}_p)_{\hat{\Omega}_f^h},$$

and define

$$\mathcal{O}(\phi^h) := \begin{pmatrix} & \mathcal{B}_2 \\ \mathcal{A} & \mathcal{B}_1 \end{pmatrix} \begin{pmatrix} \hat{\phi}_p \\ \hat{\phi}_e \end{pmatrix}.$$

Therefore we enrich the POD basis of the deformation with information of the interface variable and the POD space of the velocity with information about pressure and interface variable. More precisely we enlarge the snapshot set V_v^r to

$$V_v^s := \operatorname{span}\{\Psi^v, \mathcal{A}\Psi^p, \mathcal{B}_1\Psi^e\}.$$

The snapshot set V_u^r is enriched to

$$V_u^s := \operatorname{span}\{\Psi^u, \mathcal{B}_2\Psi^e\}.$$

The operator $\mathcal{A} : V_p \to H^1_{0,D}(\hat{\Omega}^h_f)$ is for $\hat{\psi}^p \in \Psi^p$ defined by

$$(\widehat{\operatorname{div}}(\hat{J}\hat{F}^{-1}\hat{v}), \hat{\psi}^p)_{\hat{\Omega}^h_f} = (\mathcal{A}\hat{\psi}^p, \hat{v})_{V^h(\hat{\Omega}^h_f)} = (\mathcal{A}\hat{\psi}^p, \hat{v})_{H^1_{0,D}(\hat{\Omega}^h_f)}, \quad \forall \hat{v} \in H^1_{0,D}(\hat{\Omega}^h_f).$$

The operators $\mathcal{B}_1 : V_e \to H^1_{0,D}(\hat{\Omega}^h_f)$ and $\mathcal{B}_2 : V_e \to H^1_{0,D}(\hat{\Omega}^h_s)$ are defined via

$$-(\hat{v}^h, \hat{\phi}_e)_{\hat{\Gamma}^h_I} = (\mathcal{B}_1\hat{\psi}^e, \hat{v})_{V^h(\hat{\Omega}^h_f)} = (\mathcal{B}_1\hat{\psi}^e, \hat{v})_{H^1_{0,D}(\hat{\Omega}^h_f)} \quad \forall \hat{v} \in H^1_{0,D}(\hat{\Omega}^h_f)$$

and

$$\frac{1}{\Delta t}(\hat{u}^h, \hat{\phi}^h_e)_{\hat{\Gamma}^h_I} = (\mathcal{B}_2\hat{\psi}^e, \hat{u})_{V^h(\hat{\Omega}^h_s)} = (\mathcal{B}_2\hat{\psi}^e, \hat{u})_{H^1_{0,D}(\hat{\Omega}^h_s)} \quad \forall \hat{u} \in H^1_{0,D}(\hat{\Omega}^h_s)$$

respectively.

8 Multilevel Optimization

The presented algorithm consists of three main parts, cf. [1]. A finite element model (FEM), the construction of the reduced order model (DROM) and the reduced optimization (OROM). The basic idea is to run the whole optimization on the ROM, and using the FE model to control the quality of the reduced order model. The optimization itself is a sequential quadratic programming (SQP) algorithm which we globalize by adding a trust region method.

Let $\hat{C}^h$ be a Hilbert space. With suitable assumptions on cost function and state equation we have a solution operator $\hat{c}^h \in \hat{C}^h \to \hat{y}^h(\hat{c}^h) \in \hat{Y}^h$ and can define the *reduced problem*

$$\min_{\hat{c}^h \in \hat{C}^h} \hat{\mathcal{J}}(\hat{c}^h) := \hat{\mathcal{J}}(\hat{y}^h(\hat{c}^h), \hat{c}^h) \quad \text{s.t. } \hat{c}^h \in \hat{C}^h := \{\hat{c}_h \in \hat{C}^h : (\hat{y}^h(\hat{c}^h), \hat{c}^h) \in W^h_{ad}\},$$

where $W_{ad} \subset W := \hat{Y}^h \times \hat{C}^h$ is nonempty and closed, see [5].

We need the derivative of the reduced cost functional

$$\hat{g}^h := \hat{\mathcal{J}}'(\hat{c}^h) = \hat{\mathcal{J}}_{\hat{y}^h}(\hat{y}^h(\hat{c}^h), \hat{c}^h)\hat{y}^{h\prime}(\hat{c}^h) + \hat{\mathcal{J}}_{\hat{c}^h}(\hat{y}^h(\hat{c}^h), \hat{c}^h),$$

which is calculated with help of the adjoint approach. Solving the adjoint equation

$$\mathcal{A}^h(\lambda^h) := \mathcal{S}^h_{\hat{y}^h}(\hat{y}^h(\hat{c}^h), \hat{c}^h)^*\lambda^h + \hat{\mathcal{J}}_{\hat{y}^h}(\hat{y}^h(\hat{c}^h), \hat{c}^h) = 0$$

for λ^h, the reduced gradient is given by

$$\hat{g}^h(\hat{c}^h) = \mathcal{S}_{\hat{c}^h}(\hat{y}^h(\hat{c}^h), \hat{c}^h)^*\lambda^h + \hat{\mathcal{J}}_{\hat{c}^h}(\hat{y}^h(\hat{c}^h), \hat{c}^h).$$

From the first order optimality conditions we obtain

$$\|P^h(-\hat{g}^h)\|_{\hat{C}^h},$$

with the projection

$$P^h(\hat{g}^h) := P^h_{\hat{C}^h_{ad}-\hat{c}^h}(\hat{g}^h) = \arg\min_{\tilde{g}^h \in \hat{C}^h_{ad}-\hat{c}^h} \|\tilde{g}^h - \hat{g}^h\|_{\hat{C}^h}$$

as a measure of optimality.

The same calculation can be done on the reduced order model, giving a reduced adjoint state λ^r and the reduced gradient $\hat{g}^r$.

While we derive the reduced gradient by adjoint calculation, we approximate the Hessian H with a BFGS approach, and denote it by $\tilde{H}$. We further use the error estimators

$$\eta^y := \mathcal{S}^h(\hat{y}^r), \text{ and } \eta^\lambda := \mathcal{A}^h(\lambda^r).$$

With the subscript n we denote the current iteration of the optimization.

Algorithm 1 Input: $\delta = (\delta_u, \delta_v, \delta_p, \delta_e) \in (0, 1]^4$, a number of snapshots $N_s \in \mathbb{N}$, optimality tolerance $\epsilon_{opt} > 0$, trust-region radius $\Delta_0 > 0$, $0 < \eta_1 < \eta_2 < 1$, $\eta_0 < \frac{1}{3}\eta_1$, initial guesses for the constants $\epsilon_0 < 1$, $\epsilon_0^y, \epsilon_0^\lambda, \epsilon_0^e > 0$, reduction constants $\epsilon_{red}, \Delta_{red} > 1$, constants for the criticality measurement $\epsilon_0, \epsilon_r \in (0, 1)$ and an initial guess for the control $c_0 \in C^h$.

FEM1 Compute the solution $\hat{y}_n = (\hat{u}_n^h, \hat{v}_n^h, \hat{p}_n^h, \hat{e}_n^h)$ for the control $\hat{c}_n$ of the discretized FEM state equation.

FEM2 Compute the FEM adjoint state $\lambda_n = (\lambda^h_{u,n}, \lambda^h_{v,n}, \lambda^h_{p,n}, \lambda^h_{e,n})$ and the reduced gradient $\hat{g}_n^h$ for the control $\hat{c}_n$.

FEM3 If $\|P^h(-\hat{g}_n^h)\|_{\hat{C}^h} \leq \epsilon_{opt}$ *stop*.

DROM1 Build a ROM model, i.e., calculate Ψ_u, Ψ_v, Ψ_p and Ψ_e, where δ defines the dimensions l_u, l_v, l_p and l_e. As snapshots take the first N_s time steps of the solution $\hat{y}_n$ and the adjoint states λ_n .

DROM2 Compute $\hat{y}_n^r = (\hat{u}_n^r, \hat{v}_n^r, \hat{p}_n^r, \hat{e}_n^r)$, $\lambda_n^r = (\lambda_{u,n}^r, \lambda_{v,n}^r, \lambda_{p,n}^r, \lambda_{e,n}^r)$, reduced gradient $\hat{g}_n^r$ and stationarity measure $\|P^h(-\hat{g}_n^r)\|_{\hat{C}^h}$ of the reduced model.

DROM3 Check

$$\|\hat{y}_n^h - \hat{y}_n^r\|_{\hat{Y}^h} \leq \min(\epsilon_n^y, \Delta_n),$$

$$\|\lambda_n^h - \lambda_n^r\|_{\Lambda^h} \leq \min(\epsilon_n^\lambda, \Delta_n),$$

$$\|P^h(-\hat{g}_n^r)\|_{\hat{C}^h} > \epsilon_0 \|P^h(-\hat{g}_n^h)\|_{\hat{C}^h}.$$

If one of the inequalities is not true and $\delta \neq (1, 1, 1, 1)$ enlarge δ and go to *DROM1*. Else if $N_s < N$, enlarge N_s and go to *DROM1*. If it is not possible to create a better ROM, then *stop*.
If all inequalities are fulfilled set $n^r = 0$ and go to *OROM1*.

OROM1 Set $n^r = n^r + 1$ and compute a step $s_n^{n_r}$ as an inexact solution of the SQP subproblem $(SQP)_n$, as well as a predicted reduction $\text{pred}^r := -\hat{g}_n^r s_n^{n_r} + 0.5 s_n^{n_r,T} \tilde{H}_n s_n^{n_r}$, satisfying a Cauchy decrease condition.

OROM2 For $\hat{c}_{n+1} := \hat{c}_n + s_n$, compute $\hat{y}_{n+1}^r, \lambda_{n+1}^r, \hat{g}_{n+1}^r, \|P^h(-g_{n+1}^r)\|_{C^h}$ and the actual reduction $\text{ared}^r := \hat{\mathcal{J}}(c_n) - \hat{\mathcal{J}}(c_n + s_n)$. If $n^r = 1$, save $\tilde{g}_1^r = \hat{g}_n^r$ and reset trust-region $\Delta = \Delta_0$.
If $\|P^h(-\hat{g}_{n+1}^r)\|_{\hat{C}^h} < \epsilon_{opt}$ go to *FEM1*.

OROM3 If

$$\|P^h(-\hat{g}_{n+1}^r)\|_{\hat{C}^h} \leq \epsilon_r \|P^h(-\tilde{g}_1^r)\|_{\hat{C}^h}$$

go to *FEM4* with $s_n^h = \sum_{l=0}^{n^r-1} s_n^l$.

OROM4 Compute error estimators $\eta_{n+1}^y, \eta_{n+1}^\lambda$ and, if $n^r > 1$, check

$$\epsilon_n^e \max(\eta_n^y \eta_n^\lambda, \eta_{n+1}^y \eta_{n+1}^\lambda) \leq \eta_0 \,\text{pred}^r(s_n^{n_r}).$$

If it holds true, set $\epsilon_{n+1}^y = \epsilon_n^y$, $\epsilon_{n+1}^\lambda = \epsilon_n^\lambda$, $\Delta_{n+1} = \Delta_n$ and go to *OROM5*. Else, set $\epsilon_{n+1}^y = \frac{\epsilon_n^y}{\epsilon_{red}}$, $\epsilon_{n+1}^\lambda = \frac{\epsilon_n^\lambda}{\epsilon_{red}}$. and check

$$\epsilon_n^e \eta_{n+1}^y \eta_{n+1}^\lambda \geq \eta_0 \,\text{pred}^r(s_n^{n_r}).$$

If this holds true, also reduce the trust region $\Delta_{n+1} = \frac{\Delta_n}{\Delta_{red}}$, else set $\Delta_{n+1} = \Delta_n$.

OROM5 Check

$$\eta_n^y \leq \min(\epsilon_n^y, \Delta_n) \tag{6}$$

$$\eta_n^\lambda \leq \min(\epsilon_n^\lambda, \Delta_n) \tag{7}$$

If (6) and (7) are satisfied go to *OROM6*. If (6) or (7) are not satisfied, set $\hat{c}_{n+1} = \hat{c}_n$, $s_{n+1}^h = \sum_{l=0}^{n^r-1} s_n^l$ go to *FEM4*.

OROM6 If $\frac{\mathrm{ared}^r(s_n^{n_r})}{\mathrm{pred}^r(s_n^{n_r})} \geq \eta_1$ accept the step. If $\frac{\mathrm{ared}^r(s_n^{n_r})}{\mathrm{pred}^r(s_n^{n_r})} \geq \eta_2$, set $\Delta_{n+1} = 2\Delta_n$, $\epsilon_{n+1}^y = \epsilon_0^y$ and $\epsilon_{n+1}^\lambda = \epsilon_0^\lambda$. Set $n = n+1$ and go to *OROM1*. Otherwise reject the step, set $\Delta_{n+1} = \frac{\Delta_n}{\Delta_{red}}$ and go to *OROM1*.

FEM4 Compute the solutions of the finite element system $\hat{y}_{n+1}$, λ_{n+1} and $\hat{g}_{n+1}^h$ for $\hat{c}_{n+1}$. If

$$\frac{\mathrm{ared}^{\mathrm{h}}(s_n^h)}{\sum_{l=0}^{n^r-1} \mathrm{pred}^{\mathrm{r}}(s_n^l)} \geq \eta_1 - 3\eta_0, \tag{8}$$

accept the step, set $n = n+1$, and go to *FEM3*.

FEM5 If (8) does not hold, go to *FEM4* with $n^r = n^r - 1$. If $n_r = 0$ then *stop*.

For a convergence analysis of a similar algorithm we refer to [1].

9 Numerical Results

For numerical tests, we use the setting of the benchmark problem, cf. Fig. 1, as prescribed in [11]. The domain has length $L = 1.2$ and height $H = 0.41$. The midpoint of the cylinder is at (0.2, 0.2) and its diameter is 0.1. The hyperelastic flag has length 0.35 and height 0.02 and is attached to the cylinder. The Reynolds number is 200, using the mean inflow and the diameter of the cylinder as characteristics.

We consider a finite element discretization with 9698 degrees of freedom in total. Thereof the deformation uses 678, the velocity 7748, pressure 1014, and interface 258, respectively. The mesh movement is realized with a linear operator acting on the pressure grid.

9.1 Reduced Order Model

The eigenvalues of the correlation matrices of the states drop satisfactorily, such that not many modes are needed to capture a lot of information. Since the movement of the flag is periodic for constant inflow, after a certain amount of time, we take snapshots from one period.

For example, the time step $\Delta t = 0.001$ has a period size ~368. For 368 time snapshots we see the corresponding eigenvalues for each state in Fig. 2. The number of required modes for given energy δ is shown in Table 1.

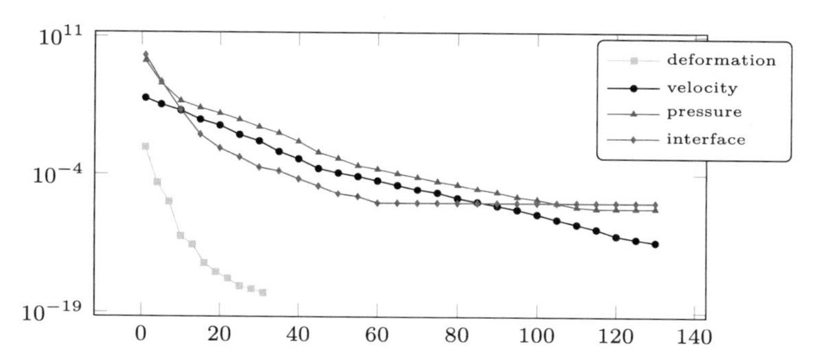

Fig. 2 Eigenvalues of the state correlation matrices

Table 1 Modes required to capture energy δ

δ	$1 - 10^{-5}$	$1 - 10^{-6}$	$1 - 10^{-7}$	$1 - 10^{-8}$	$1 - 10^{-9}$	$1 - 10^{-10}$
l_u	5	7	8	8	9	11
l_v	31	36	42	52	63	74
l_p	18	25	32	37	42	50
l_e	8	10	11	14	17	21

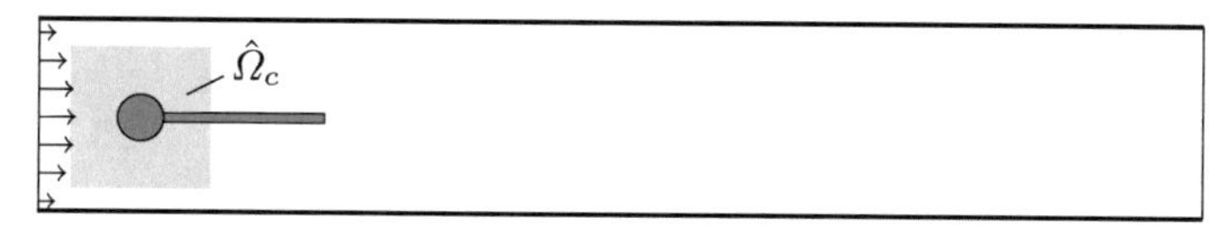

Fig. 3 Control domain $\hat{\Omega}_c$ for C_1^h

9.2 Optimization

As a test case, we consider the minimization of the lift forces acting on the structure and the cylinder:

$$\hat{\mathcal{J}}(\hat{c}^h) := \hat{\mathcal{J}}(\hat{y}^h, \hat{c}^h) = \int_0^T \int_{\hat{\Gamma}_I \cup \hat{\Gamma}_C} lift(\hat{y}^h, \hat{c}^h)^2 \, d\hat{x} \, dt + \alpha \int_0^T \|c^h\|_{C_h}^2 \, dt,$$

where $\hat{\Gamma}_C$ denotes the boundary between cylinder and fluid. We add a regularity term for the control in the corresponding space, with constant $\alpha > 0$.

We will consider two types of controls. In the first one the control is a body force, acting on a squared domain $\hat{\Omega}_c$ around the cylinder, see Fig. 3, with $\hat{\Omega}_c \cap \delta\hat{\Omega}_f = \emptyset$. Then, let $C_1^h = L^h(\hat{\Omega}_c)$, and $C_{1,ad}^h := \{c^h \in C_1^h, c_l \leq c^h \leq c_u\}$, with lower and upper bounds c_l, c_u. The second control is given by additional in- resp. outflows on the top and bottom wall, cf. Fig. 4. This was proposed by [9]. We prescribe at the

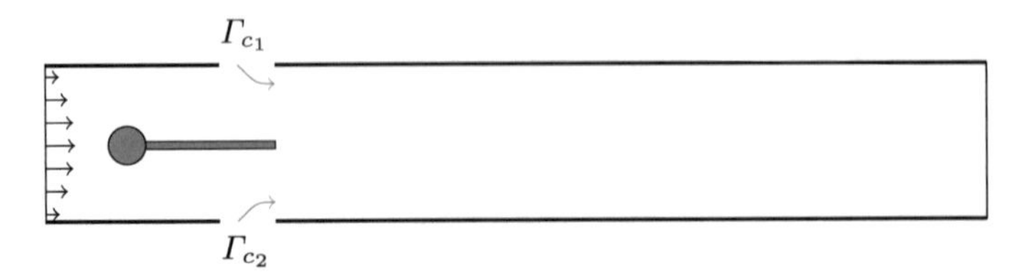

Fig. 4 Additional inflows at Γ_{c_1}, Γ_{c_2} for C_2^h

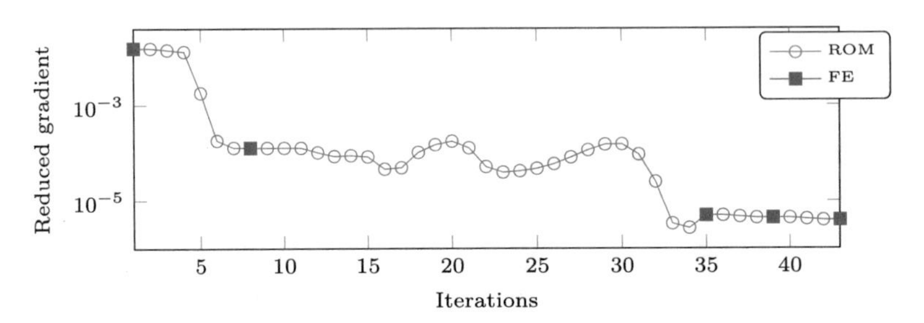

Fig. 5 Norm of the reduced gradient for $\hat{\mathcal{J}}$, $C_{1,ad}^h$

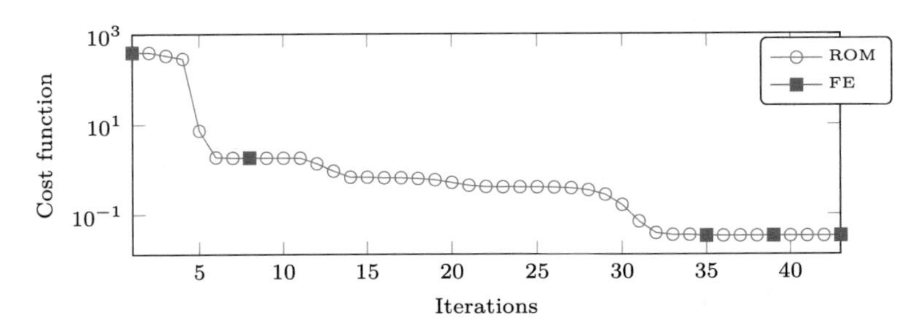

Fig. 6 Cost function $\hat{\mathcal{J}}$ under control $C_{1,ad}^h$

boundaries Γ_{c_1}, Γ_{c_1} the Dirichlet conditions

$$v_c^h(\hat{x}) := \begin{cases} c_1(\hat{x} - 0.45)(\hat{x} - 0.6), & \text{on } \Gamma_{c_1}, \\ c_2(\hat{x} - 0.45)(\hat{x} - 0.6), & \text{on } \Gamma_{c_2}. \end{cases}$$

The control is then given by $c^h := (c_1, c_2) \in \mathbb{R}^2$, and $C_{2,ad}^h := \{(c_1, c_2) \in \mathbb{R}^2, c_l \leq c_1, c_2 \leq c_u\}$, with box constraints $c_l < c_u \in \mathbb{R}$.

In Figs. 5 and 6 we see the development of reduced gradient and cost function $\hat{\mathcal{J}}$ under the control of the body force $C_{1,ad}^h$, respectively. We see, that we only need 5 evaluations of the finite element model to reach $\epsilon_{opt} := 1e^{-05}$. If we plot the lift for the calculated optimal control we see that it is almost zero, cf. Fig. 7. This is due

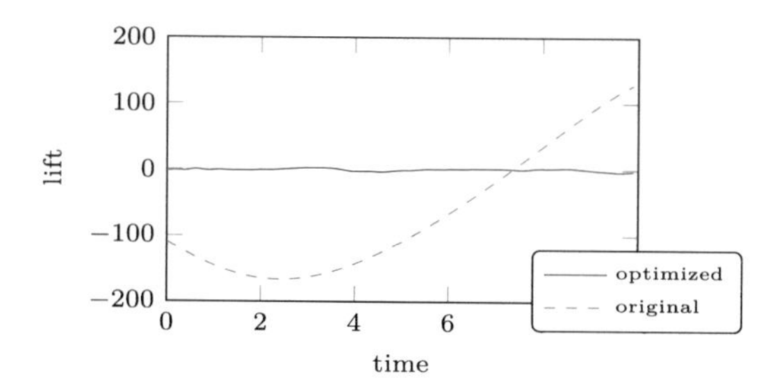

Fig. 7 Optimal controlled lift for C_1^h, 100 time steps

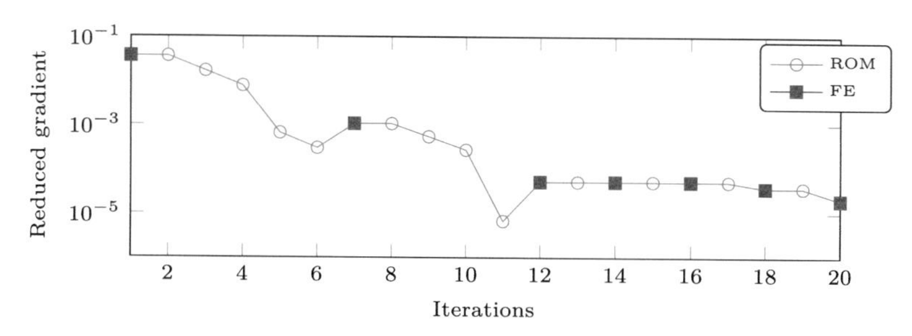

Fig. 8 Norm of the reduced gradient for $\hat{\mathcal{J}}$, $C_{2,ad}^h$

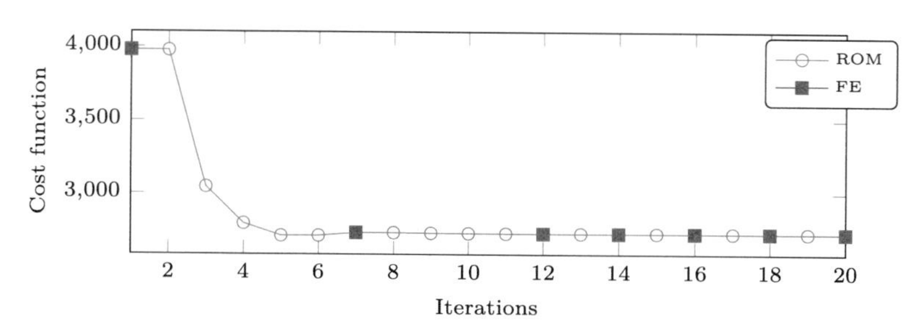

Fig. 9 Cost function $\hat{\mathcal{J}}$ under control $C_{2,ad}^h$

to the fact that the flow field around the cylinder responses immediately to the body force. For the second control we see in Figs. 8 and 9 that after only three evaluations of the finite element model, the projected gradient has dropped to $3e^{-05}$. Afterwards the reduced order model can not capture the changes of the state very good and has to be updated after every step of the optimization. The optimized lift is shown in Fig. 10. Here we see the damping resulting of the additional inflows.

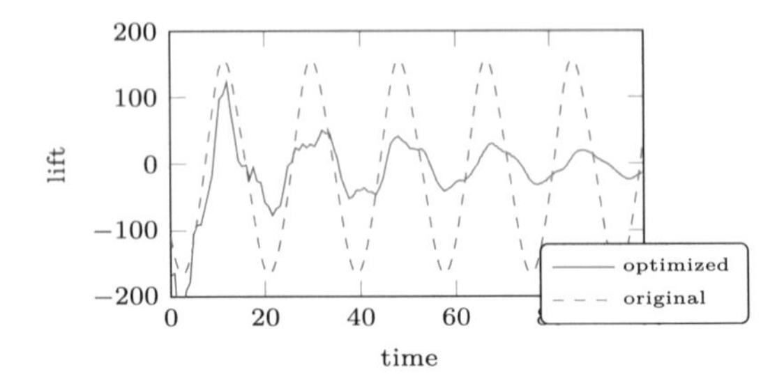

Fig. 10 Optimal controlled lift for C_2^h, 1000 time steps

Acknowledgements This work is supported by the 'Excellence Initiative' of the German Federal and State Governments and the Graduate School of Computational Engineering at Technische Universität Darmstadt.

References

1. Bott, S.: Adaptive SQP method with reduced order models for optimal control problems with constraints on the state applied to the Navier-Stokes equations (Diss.). Dr. Hut, München; TU Darmstadt, Fachbereich Mathematik, Darmstadt (2016)
2. Donea, J.: Arbitrary Lagrangian-Eulerian finite element methods. In: Computational Methods for Transient Analysis, Mech. Math. Methods, 1. Ser., Comput. Methods Mech. 1, pp. 473–516 (1983)
3. Ghiglieri, J.: Optimal flow control based on POD and MPC and an application to the cancellation of Tollmien-Schlichting waves (Diss.). Dr. Hut, München; TU Darmstadt, Fachbereich Mathematik, Darmstadt (2014)
4. Hinze, M., Volkwein, S.: Error estimates for abstract linear-quadratic optimal control problems using proper orthogonal decomposition. Comput. Optim. Appl. **39**(3), 319–345 (2008)
5. Hinze, M., Pinnau, R., Ulbrich, M., Ulbrich, S.: Optimization with PDE Constraints. Springer, Dordrecht (2009)
6. Hron, J., Turek, S.: A monolithic FEM/multigrid solver for an ALE formulation of fluid-structure interaction with applications in biomechanics. In: Fluid-Structure Interaction. Lecture Notes in Computational Science and Engineering, vol. 53, pp. 146–170. Springer, Berlin (2006)
7. Kunisch, K. Volkwein, S.: Proper orthogonal decomposition methods for a general equation in fluid dynamics. SIAM J. Numer. Anal. **40**(2), 492–515 (2002)
8. Raymond, J.-P., Vanninathan, M.: A fluid-structure model coupling the Navier-Stokes equations and the Lamé system. J. Math. Pures Appl. (9) **102**(3), 546–596 (2014)
9. Razzaq, M., Tsotskas, C., Turek, S., Kipouros, T., Savill, M., Hron, J.: Multi-objective optimization of a fluid structure interaction benchmarking. Comput. Model. Eng. Sci. **90**(4), 303–337 (2013)
10. Rozza, G., Veroy, K.: On the stability of the reduced basis method for Stokes equations in parametrized domains. Comput. Methods Appl. Mech. Eng. **196**(7), 1244–1260 (2007)
11. Turek, S., Hron, J.: Proposal for numerical benchmarking of fluid-structure interaction between an elastic object and laminar incompressible flow. In: Fluid-Structure Interaction. Modelling, Simulation, Optimisation. Proceedings of the Workshop, Hohenwart, October 2005, pp. 371–385. Springer, Berlin (2006)
12. Volkwein, S.: Optimal control of a phase-field model using proper orthogonal decomposition. J. Appl. Math. Mech./Z. Angew. Math. Mech. **81**(2), 83–97 (2001)

An Investigation of Airfoil Noise Prediction Using Hybrid LES/RANS Models

Xin Huang and Michael Schäfer

Abstract Within the framework of this study, two hybrid LES/RANS turbulence models, namely the limited numerical scales model (LNS) and the very large eddy simulation model (VLES), are investigated for the aeroacoustic simulation based on a NACA0012 test case. The angle of attack is set to 0°, 10.8° and 14.4°. The simulation results are compared with that of the well-established large eddy simulation (LES) model.

Keywords Aeroacoustics · Limited numerical simulation · Very large eddy simulation

1 Introduction

In recent decades, the study of aerodynamic noise for engineering applications has gained much attention. The aeroacoustic quantities can be computed by solving the compressible Navier-Stokes equations directly. This method, often referred to as Direct Numerical Simulation (DNS) [1], is not widely applied due to the lack of efficiency, considering that the length scales as well as time scales of flow and acoustic fields lie obviously far to each other. The expansion about incompressible flow technique (EIF) proposed by Hardin and Pope [6] and modified by Shen and

X. Huang (✉)
Graduate School of Computational Engineering, Technische Universität Darmstadt, Darmstadt, Germany

Institute of Numerical Methods in Mechanical Engineering, Technische Universität Darmstadt, Darmstadt, Germany
e-mail: huang@gsc.tu-darmstadt.de

M. Schäfer
Graduate School of Computational Engineering, Technische Universität Darmstadt, Darmstadt, Germany

M. Schäfer et al. (eds.), *Recent Advances in Computational Engineering*, Lecture Notes in Computational Science and Engineering 124,
https://doi.org/10.1007/978-3-319-93891-2_3

Sørensen [12, 13] separates the computation of the flow and the acoustic field and significantly reduces the computational efforts.

For turbulent flows, a turbulence model can be adopted to characterize the unresolved turbulence scale. Reynolds Averaged Navier-Stokes equations model (RANS) is computationally advantageous, however, it is not able to capture the transient small scale structures, which contribute a considerable part to the noise generation, because only the time averaged flow is resolved. In the large eddy simulation (LES), the large scale structures are resolved while the small scale structures are implicitly accounted for by using a sub-grid scale model, making it able to predict certain amount of anisotropic turbulent structures. However, the LES model is almost only used in academic research due to the high computational cost required for engineering problems. Consequently, hybrid LES/RANS models combining the advantages of the LES model and the RANS model become of interest.

This paper presents the results of aeroacoustic simulations using hybrid turbulence models. The flow is simulated using two different hybrid LES/RANS methods: very large eddy simulation (VLES) and limited numerical scales (LNS), which are compared with the large eddy simulation (LES).

2 Linearized Euler Equations for Aeroacoustic Simulation

The expansion about incompressible flow technique (EIF) is utilized to decompose the flow and the acoustic field. According to this assumption, the compressible flow field can be divided into an incompressible flow in the background and acoustic fluctuations

$$u_i = u_i^{\text{inc}} + u_i^{\text{ac}}, \tag{1}$$

$$p_i = p_i^{\text{inc}} + p_i^{\text{ac}}, \tag{2}$$

$$\rho_i = \rho_i^{\text{inc}} + \rho_i^{\text{ac}}, \tag{3}$$

where u_i, p_i and ρ_i are the velocity, pressure and density of compressible flow and superscripts inc and ac denote the components of incompressible flow and acoustic field, respectively. Instead of using the acoustic pressure directly, the sound pressure level (SPL) L_p with unit decibel (dB) is usually used to describe the sound value:

$$L_p = 10 \cdot \lg \frac{\left(p_{eff}^{\text{ac}}\right)^2}{p_{ref}^2} = 20 \cdot \lg \frac{p_{eff}^{\text{ac}}}{p_{ref}}, \tag{4}$$

where p_{ref} is a reference acoustic pressure and p^{ac}_{eff} is an effective acoustic pressure given by

$$p^{ac}_{eff} = \sqrt{\frac{1}{T}\int_{t=0}^{T} (p^{ac}(t))^2 \, dt}. \tag{5}$$

The acoustic pressure p^{ac}, the acoustic density fluctuation ρ^{ac} and the particle velocity u_i^{ac} are governed by the linearized Euler equations (LEE), given as [9]

$$\frac{\partial \rho^{ac}}{\partial t} + \rho^{inc}\frac{\partial u_i^{ac}}{\partial x_i} + u_i^{inc}\frac{\partial \rho^{ac}}{\partial x_i} = 0, \tag{6}$$

$$\rho^{inc}\frac{\partial u_i^{ac}}{\partial t} + \rho^{inc} u_j^{inc}\frac{\partial u_i^{ac}}{\partial x_j} + \frac{\partial p^{ac}}{\partial x_i} = 0, \tag{7}$$

$$\frac{\partial p^{ac}}{\partial t} + c^2\rho^{inc}\frac{\partial u_i^{ac}}{\partial x_i} + c^2 u_i^{inc}\frac{\partial \rho^{ac}}{\partial x_i} = -\frac{\partial p^{inc}}{\partial t}. \tag{8}$$

3 Hybrid LES/RANS Methods

In general, the motion of flow is governed by the Navier-Stokes equations, given as

$$\frac{\partial u_i^{inc}}{\partial x_i} = 0, \tag{9}$$

$$\frac{\partial u_i^{inc}}{\partial t} + \frac{\partial u_i^{inc} u_j^{inc}}{\partial x_j} = \frac{\partial}{\partial x_j}\left(\nu \frac{\partial u_i^{inc}}{\partial x_j}\right) - \frac{1}{\rho^{inc}}\frac{\partial p^{inc}}{\partial x_i}. \tag{10}$$

According to the Reynolds decomposition, the unsteady flow motion can be seen as superposition of a mean flow and a fluctuation

$$\phi(x_i, t) = \overline{\phi(x_i, t)} + \phi'(x_i, t), \tag{11}$$

where $\overline{\phi(x_i, t)}$ is the ensemble or time averaged part and $\phi'(x_i, t)$ is the fluctuation component. Inserting the Reynolds decomposition into Navier-Stokes equations yields equations governing the mean flow motion, called Reynolds Averaged Navier-Stokes (RANS)

$$\frac{\partial \overline{u_i^{inc}}}{\partial x_i} = 0, \tag{12}$$

$$\frac{\partial \overline{u_i^{inc}}}{\partial t} + \frac{\partial \overline{u_i^{inc}}\,\overline{u_j^{inc}}}{\partial x_j} = \frac{\partial}{\partial x_j}\left(\nu \frac{\partial \overline{u_i^{inc}}}{\partial x_j} - \overline{u_i' u_j'}\right) - \frac{1}{\rho^{inc}}\frac{\partial \overline{p^{inc}}}{\partial x_i}. \tag{13}$$

The additional unknown term $\overline{u'_i u'_j}$, often referred to as Reynolds stress tensor, can be modeled by the Boussinesq approximation

$$\overline{u'_i u'_j} = -2\nu_t S_{ij} + \frac{2}{3}k\delta_{ij} \quad \text{with} \quad S_{ij} = \frac{1}{2}\left(\frac{\partial \overline{u_i^{\mathrm{inc}}}}{\partial x_j} + \frac{\partial \overline{u_j^{\mathrm{inc}}}}{\partial x_i}\right), \tag{14}$$

with the turbulent viscosity ν_t, the turbulent kinetic energy k and the strain rate tensor S_{ij}.

3.1 Very Large Eddy Simulation

Very large eddy simulation, also called flow simulation methodology in some literatures, was first proposed by Speziale [14]. The original idea was to damp the Reynolds stress tensor $\overline{u'_i u'_j}^{(R)}$ obtained from conventional RANS models via

$$\overline{u'_i u'_j} = F_r \overline{u'_i u'_j}^{(R)}, \tag{15}$$

with F_r being a coefficient, which is defined as

$$F_r = [1 - \exp(-\beta L^{\Delta}/L^k)]^n, \tag{16}$$

where L^{Δ} denotes the filter width correlated with the computational mesh size and L^k is the Kolmogorov length scale

$$L^k = \left(\frac{\nu^3}{\varepsilon}\right)^{1/4}. \tag{17}$$

As L^{Δ}/L^k approaches 0 the Reynolds stress tensor vanishes, leading to a direct numerical simulation for the turbulent flow. On the contrary, if L^{Δ}/L^k approaches ∞ we have a regular RANS model. However, the original model has some shortcomings. First of all, the parameters β and n in Eq. (16) were not defined by Speziale. Furthermore, it is not necessary to set the Kolmogorov length scale as the limit to ensure a LES behaviour between the limits [15]. In addition, the original VLES model tends to overdamp the Reynolds stress tensor in the near wall region [8]. Therefore, different formulations of the damping coefficient have been proposed. In this paper, we apply the modification proposed by Chang et al. [4], where F_r is derived based on the relation

$$F_r \propto \int_{\kappa_{\mathrm{C}}(\mathrm{LES})}^{\kappa_{\mathrm{K}}} E(\kappa)\,\mathrm{d}\kappa \Bigg/ \int_{\kappa_{\mathrm{C}}(\mathrm{VLES})}^{\kappa_{\mathrm{K}}} E(\kappa)\,\mathrm{d}\kappa, \tag{18}$$

where κ_{K} is the wave number of Kolmogorov scale, and $\kappa_{\mathrm{C}}(\mathrm{LES})$ and $\kappa_{\mathrm{C}}(\mathrm{VLES})$ are the cut-off wave number of the VLES model and the LES model, respectively. Based on this assumption, F_r is formulated as

$$F_r = \min\left[\left(\frac{L^{\Delta}}{k^{3/2}/\varepsilon}\right)^{\frac{4}{3}}, 1\right]. \tag{19}$$

3.2 Limited Numerical Scales

The limited numerical scales, proposed by Batten et al. [2, 3], is based on the same idea as the very large eddy simulation. The function F_r, named as latency factor by Batten, has the following form [5]

$$F_r = \frac{\min(\nu_{\mathrm{t}}^{\mathrm{LES}}, \nu_{\mathrm{t}}^{\mathrm{RANS}})}{\nu_{\mathrm{t}}^{\mathrm{RANS}}}, \tag{20}$$

where $\nu_{\mathrm{t}}^{\mathrm{RANS}}$ is the turbulent viscosity obtained by an underlying RANS model, and $\nu_{\mathrm{t}}^{\mathrm{LES}}$ is the turbulent viscosity of an LES model. In the present work, we select the Smagorinsky model as the LES model.

When the grid is sufficiently fine, the LES branch $\nu_{\mathrm{t}}^{\mathrm{LES}}$ is chosen and the turbulent viscosity is scaled down to LES-like values. In the contrary, when the grid is coarse, the RANS branch $\nu_{\mathrm{t}}^{\mathrm{RANS}}$ is applied and the turbulent viscosity is modeled using the underlying RANS model. In the following, the Chien $k-\varepsilon$ based LNS model is presented. The equations are derived based on (20).

The governing equations of the fundamental Chien $k-\varepsilon$ RANS model are given as

$$\rho\frac{\partial k}{\partial t} + \rho u_j\frac{\partial k}{\partial x_j} = \frac{\partial}{\partial x_j}\left[\left(\mu + \frac{\mu_{\mathrm{t}}}{\sigma_k}\right)\frac{\partial k}{\partial x_j}\right] + P_k - \rho\varepsilon - 2\nu\frac{k}{y^2}, \tag{21}$$

$$\rho\frac{\partial \varepsilon}{\partial t} + \rho u_j\frac{\partial \varepsilon}{\partial x_j} = \frac{\partial}{\partial x_j}\left[\left(\nu + \frac{\nu_{\mathrm{t}}}{\sigma_\varepsilon}\right)\frac{\partial \varepsilon}{\partial x_j}\right] + C_{\varepsilon 1}P_k\frac{\varepsilon}{k} - \varepsilon\rho\left(\frac{C_{\varepsilon 2}f_\varepsilon}{k} - \frac{2\nu}{y^2}e^{-0.5y^+}\right), \tag{22}$$

with the production term

$$P_k = 2\mu_{\mathrm{t}}S_{ij}S_{ij}, \quad \text{with} \quad S_{ij} = \frac{1}{2}\left(\frac{\partial u_i}{\partial x_j} + \frac{\partial u_j}{\partial x_i}\right), \tag{23}$$

and the damping function

$$f_\varepsilon = 1 - 0.22e^{-Re_T/6}, \quad \text{with} \quad Re_T = \frac{\rho k^2}{\mu\varepsilon}. \tag{24}$$

Table 1 Model constants for LNS model

C_μ	$C_{\varepsilon 1}$	$C_{\varepsilon 2}$	σ_k	σ_ε
0.09	1.35	1.80	1.0	1.3

In the LNS model, the turbulent viscosity is calculated via

$$\nu_\mathrm{t} = F_r C_\mu f_\mu k^2/\varepsilon, \quad \text{with} \quad f_\mu = 1 - e^{-0.0115 y^+}. \tag{25}$$

F_r is formulated as

$$F_r = \min\left\{ \frac{(C_\mathrm{s}\Delta)^2 \sqrt{2 S_{ij} S_{ij}}}{C_\mu f_\mu k^2/\varepsilon}, 1 \right\}. \tag{26}$$

The coefficients used for this model are listed in Table 1.

4 Numerical Methods

The flow domain is discretized using the finite volume method with a structured grid. In order to perform a parallel computation, the grid is divided into multiple blocks, distributed to different processors during the computation. The communication between different blocks is realized using ghost cells, which are additional fictional cells adjacent to the block interfaces [11]. In terms of time discretization, a second order implicit Euler method is applied. The diffusive flux is approximated using central differencing scheme, while the convective flux is obtained from a blending between the CDS and GAMMA schemes. The pressure and velocity are coupled using the SIMPLE algorithm. After solving the flow domain, the acoustic source term is transfered to the acoustic solver, as shown in Fig. 1.

The acoustic sources are interpolated onto the acoustic acoustic grid, which has a relatively coarse resolution in comparison to the flow grid, using a trilinear interpolation. The linearized Euler equations are solved using a high-resolution scheme with a flux limiter of van Leer. The acoustic solver employs an explicit Euler method to discretize the time domain. Consequently, the Courant number should always fulfill the CFL criterion

$$\mathrm{CFL}_\mathrm{CAA} = \frac{c \Delta t_\mathrm{CAA}}{\Delta x} < 1, \tag{27}$$

where c is the sound speed, Δt_CAA is the time step of the acoustic computation and Δx is the cell size. Hence, a frozen fluid method is applied, where the acoustic computation is conducted multiple times for each time step of the flow. For the DDES and the LNS simulation, the filter width is set to the maximum of the cell

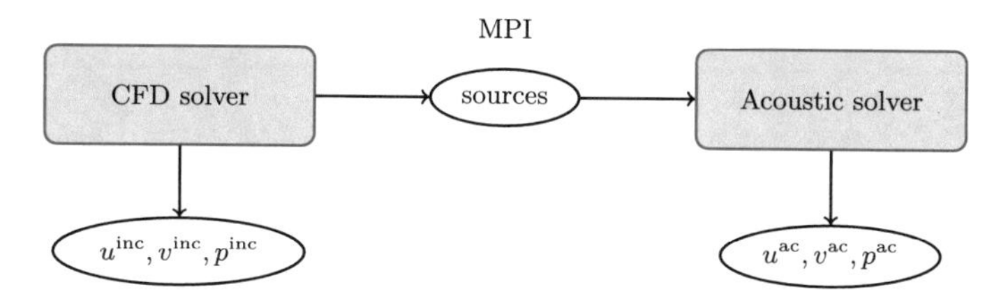

Fig. 1 Numerical realization of flow solver and acoustic solver

size

$$\Delta_{\max} = \max(\Delta_x, \Delta_y, \Delta_z), \tag{28}$$

while the VLES model has a filter width of

$$\Delta = \min\{\max(C_w d_w, C_w h_{\max}, h_{wn}), h_{\max}\}, \tag{29}$$

where d_w represents the distance to the wall, h_{wn} is the grid size in the wall-normal direction, $h_{\max}$ is a maximum local grid spacing, and C_w is a constant set to 0.15 [8].

5 Results and Discussion

In order to investigate different hybrid turbulence models in terms of acoustic simulation for attached flow and separated flow, the airfoil NACA0012 is simulated at three different angles of attack (AOA): 0°, 10.8° and 14.4°. The flow velocity is 39.6 m/s, corresponding to a Mach number of 0.12, fulfilling the incompressibility assumption. The Reynolds number based on a chord length of 2.54 cm is approximately 6.8×10^4.

Two different O-type structured grids are used for the LES simulation and the hybrid LES/RANS simulation. The grid configurations for three simulations are listed in Table 2. For the LES simulation, the streamwise, spanwise and wall normal spacings fulfill the requirement $\Delta x^+ = 60$, $\Delta z^+ = 20$ and $\Delta y^+ = 1$ [7], resulting in a total number of CVs of around 5 million. The grid of the flow domain for the VLES simulation and the LNS simulation has totally 1.6 million CVs, since the grid is coarsened such that $\Delta x^+ = 200$, $\Delta z^+ = 60$ and $\Delta y^+ = 2$. The same acoustic grid with 0.34 million CVs is employed for all turbulence models.

Figure 2 depicts the blending function F_r for the LNS model. It can be seen that the fluid domain is dominated by F_r close to 0, resulting in an LES simulation or even a quasi DNS simulation. In the boundary layer the model operates in a RANS mode, where F_r is increased up to 1. For the cases where flow separation occurs, the wake region is resolved by a mixture of RANS and LES modes.

Table 2 Grid resolution for LES and hybrid RANS/LES models

Simulation	Turbulence model	Δx^+	Δy^+	Δz^+	CVs for flow domain	CVs for acoustic domain
No. 1	LES	60	1	20	5 million	0.34 million
No. 2	VLES	200	2	60	1.6 million	0.34 million
No. 3	LNS	200	2	60	1.6 million	0.34 million

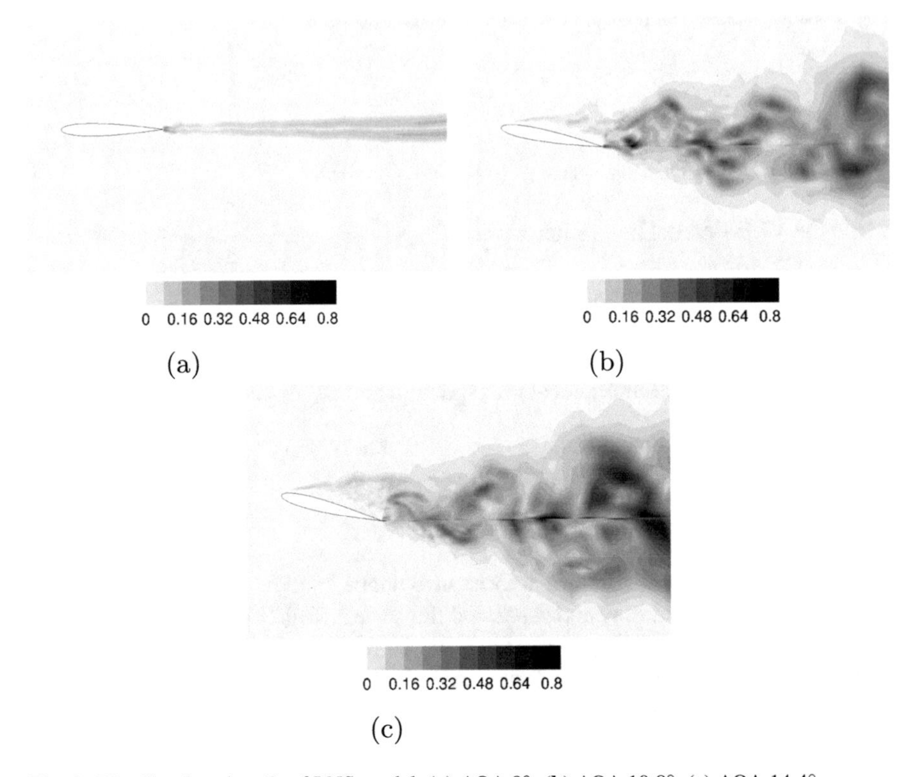

Fig. 2 Blending function F_r of LNS model. (**a**) AOA 0°. (**b**) AOA 10.8°. (**c**) AOA 14.4°

Figure 3 shows the magnitude of the acoustic sources at AOA=0°. As can be seen from the LES simulation, the largest values appear close to the trailing edge. The wake region farther from the edge have little to no influence to the source term. This phenomenon has also been observed in other research, e.g. [16]. In comparison to the LES model, the LNS and the VLES models seem to underestimate the source term in the trailing edge region and overestimate the source term on the airfoil surface. The application of the RANS mode in the boundary layer and wake region could be the reason leading to the misestimation in these regions.

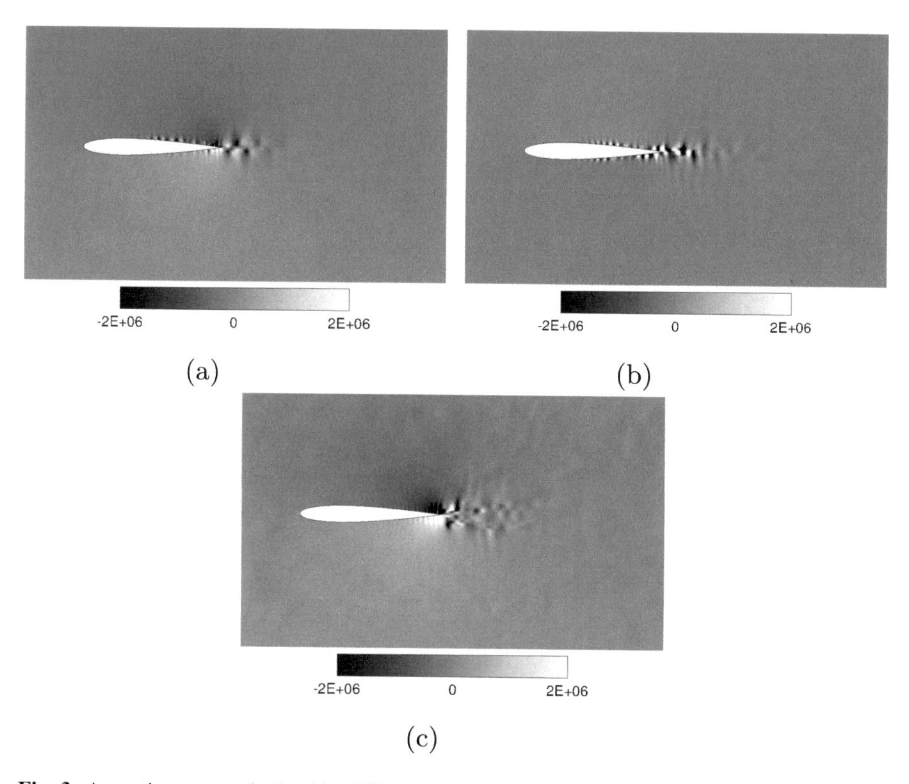

Fig. 3 Acoustic sources obtained by different turbulence models at AOA=0°. (**a**) LNS. (**b**) VLES. (**c**) LES

When the AOA is increased to 10.8° as shown in Fig. 4, a flow separation can be observed by the LES and LNS models. According to [10], the separation of turbulent flow with low to moderate Reynolds number (5.5×10^4–2.1×10^5) occurs at approximately 9.25°, and a full separation can be achieved when AOA is larger than 12°. Figure 5 illustrates that the LNS model is able to predict the separation correctly. However, the small flow structures observed in the LES simulation are merged to larger structures. The VLES model fails to predict the separation, leading to an underprediction of the acoustic sources. When the AOA reaches 14.4°, all three turbulence models succeeded to predict the separation, as can be seen from Fig. 5. The acoustic sources obtained from the VLES model shows more small scale structures while the magnitudes of the acoustic sources are similar for both models.

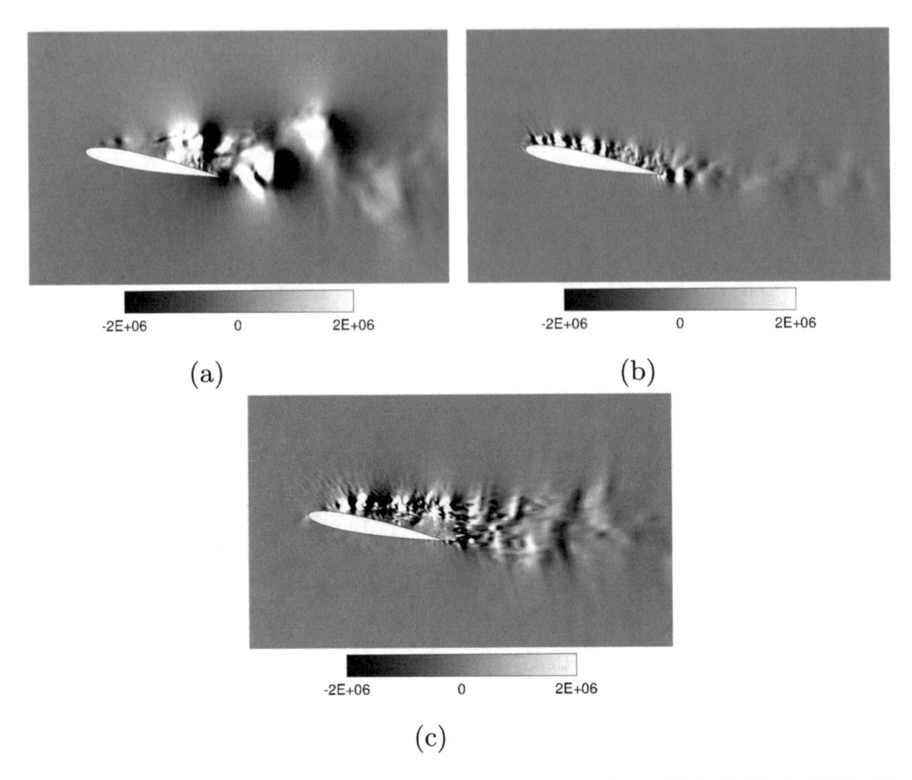

Fig. 4 Acoustic sources obtained by different turbulence models at AOA=10.8°. (**a**) LNS. (**b**) VLES. (**c**) LES

In order to obtain the directivity characteristics of the results, the acoustic pressure at 36 different points are collected, which have a distance of 1.22 m to the leading edge of the airfoil and are equidistantly distributed on a circle. Performing a Fourier Transformation on the data leading to a spectrum for the data of each point. Comparing the spectrum of all the points at the fundamental frequency, the directivity characteristics are obtained. Figure 6 displays the directivity characteristics predicted by different turbulence models. When the airfoil is set completely horizontal (AOA=0°), a dipole profile can be observed for all three turbulence models. In particular, the LNS model shows better agreement with the LES result than the VLES model. When the AOA is changed to 10.8°, where the separation of flow starts to occur, superior result is obtained by the LNS model due to the better prediction of the separation. When the AOA is increased to 14.4°, where the flow is fully separated, both the LNS and VLES models are able to deliver similar results compared to the LES model.

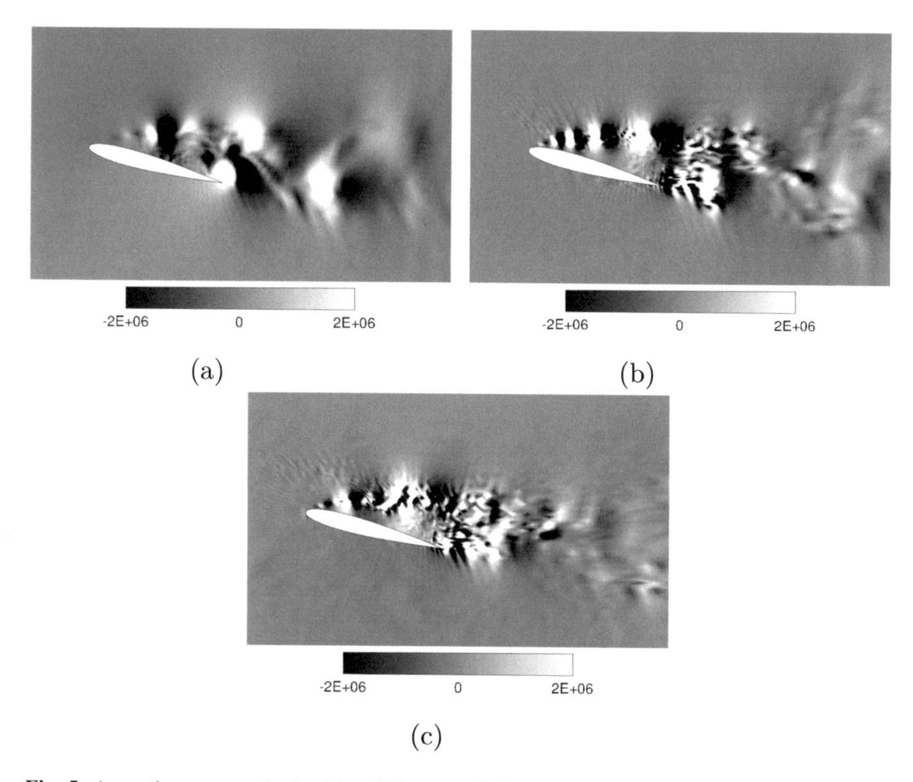

Fig. 5 Acoustic sources obtained by different turbulence models at AOA=14.4°. (**a**) LNS. (**b**) VLES. (**c**) LES

6 Conclusion

In this paper, the aeroacoustic performance of two hybrid turbulence models, the LNS model and the VLES model, were scrutinized based on NACA0012 test cases by comparison with the LES model, which serves as reference data. For fully attached flow(AOA=0°), the LNS model delivers slightly better results, although the acoustic sources obtained from the LNS model and the VLES model are similar. The LNS model achieves better prediction when the flow separation just starts to occur (AOA=10.8°). For fully separated flow (AOA=14.4°), both the LNS and the VLES models achieve satisfactory results, although the VLES model provides more details of small scale structures in the acoustic sources.

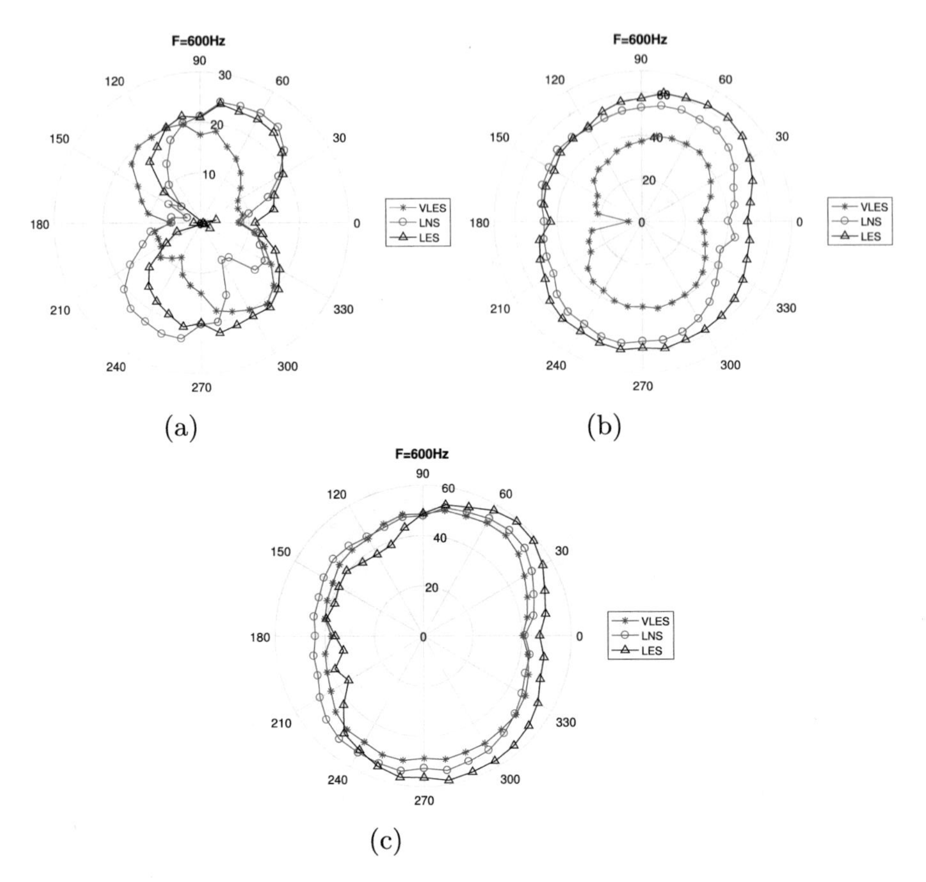

Fig. 6 Directivity characteristics of acoustic simulation results. (**a**) AOA=0°. (**b**) AOA=10.8°. (**c**) AOA=14.4°

Acknowledgements This work is supported by the 'Excellence Initiative' of the German Federal and State Governments and the Graduate School of Computational Engineering at Technische Universität Darmstadt.

References

1. Bailly, C., Bogey, C., Marsden, O.: Progress in direct noise computation. Int. J. Aeroacoust. **9**(1), 123–143 (2010)
2. Batten, P., Goldberg, U., Chakravarthy, S.: LNS - an approach towards embedded LES. In: 40th AIAA Aerospace Sciences Meeting & Exhibit, Reno, NV (2002)
3. Batten, P., Goldberg, U., Chakravarthy, S.: Interfacing statistical turbulence closures with large-eddy simulation. AIAA J. **42**(3), 485–492 (2004)
4. Chang, C.-Y., Jakirlić, S.: Swirling flow in a tube with varialby-shaped outlet orifices: an LES and VLES study. Int. J. Heat Fluid Flow **49**, 28–42 (2014)

5. Fröhlich, J., Terzi, D.: Hybrid LES/RANS methods for the simulation of turbulent flows. Prog. Aerosp. Sci. **44**(2008), 349–377 (2008)
6. Hardin, J.C., Pope, D.S.: An acoustic/viscous splitting technique for computational aeroacoustics. Theor. Comput. Fluid Dyn. **6**(5), 323–340 (1994)
7. Kaltenbach, H., Choi, H.: Large-eddy simulation of flow around an airfoil on a structured mesh. Center for Turbulence Research Annual Research Briefs, pp. 51–60 (1995)
8. Kondratyuk, A.: Investigation of the very large eddy simulation model in the context of fluid-structure interaction. Dissertation, TU Darmstadt (2017)
9. Kornhaas, M., Schäfer, M., Sternel, D.C.: Efficient numerical simulation of aeroacoustics for low Mach number flows interacting with structures. Comput. Mech. **55**, 1143–1154 (2015)
10. Rodríguez, I., Lehmkuhl, O., Borrell R., Oliva, A.: Direct numerical simulation of a NACA0012 in full stall. Int. J. Heat Fluid Flow **43**, 194–302 (2013)
11. Schäfer, M.: Computational Engineering - Introduction to Numerical Methods. Springer, Berlin (2006)
12. Shen, W.Z., Sörensen, J.N.: Aeroacoustic modelling of low-speed flows. Theor. Comput. Fluid Dyn. **13**(4), 271–289 (1999)
13. Shen, W.Z., Sörensen, J.N.: Comment on the aeroacoustic formulation of Hardin and Pope. AIAA J. **37**(1), 141–143 (1999)
14. Speziale, C.: Turbulence modeling for time-dependent RANS and VLES. AIAA J. **36**(2), 173–184 (1998)
15. Wagner, C.A., Hüttl, T., Sagaut, P.: Large-Eddy Simulation for Acoustics. Cambridge University Press, Cambridge (2007)
16. Wang, M.: Computation of trailing-edge noise at low Mach number using LES and acoustic analogy. Center for Turbulence Research Annual Research Briefs, pp. 91–106 (1998)

A High-Order Local Discontinuous Galerkin Scheme for Viscoelastic Fluid Flow

Anne Kikker and Florian Kummer

Abstract Coping with the so called high Weissenberg number problem (HWNP) is a key focus of research in computational rheology. By numerically simulating viscoelastic flow a breakdown in convergence often occurs for different computational approaches at critically high values of the Weissenberg number. This is due to two major problems concerning stability in the discretization. First, we have a mixed hyperbolic-elliptic problem weighted by a ratio parameter between retardation and relaxation time of viscoelastic fluid. Second, we have a convection-dominated convection-diffusion problem in the constitutive equations. We introduce a solver for viscoelastic Oldroyd B flow with an exclusively high-order Discontinuous Galerkin (DG) scheme for all equations using a local DG formulation in order to solve the hyperbolic constitutive equations and using a streamline upwinding formulation for the convective fluxes of the constitutive equations. The successful implementation of the local DG formulation for Newtonian fluid with appropriate fluxes containing stabilizing penalty parameters is shown in two results. First, a h^k-convergence study is presented for a non-polynomial manufactured solution for the Stokes system. Second, numerical results are shown for the confined cylinder benchmark problem for Navier-Stokes flow and compared to the same flow using a symmetric interior penalty method without additional constitutive equations.

Keywords Local discontinuous Galerkin method · Oldroyd B fluid

---

A. Kikker (✉)
Technische Universität Darmstadt, Darmstadt, Germany

Graduate School CE, Technische Universität Darmstadt, Darmstadt, Germany
e-mail: kikker@fdy.tu-darmstadt.de

F. Kummer
Technische Universität Darmstadt, Darmstadt, Germany

M. Schäfer et al. (eds.), *Recent Advances in Computational Engineering*, Lecture Notes in Computational Science and Engineering 124,
https://doi.org/10.1007/978-3-319-93891-2_4

1 Introduction

By numerically simulating viscoelastic flow a breakdown in convergence often occurs for different computational approaches at critically high values of the Weissenberg number $\mathrm{We} = \lambda_1 \frac{U}{L}$ with a characteristic velocity U, a characteristic length L and the relaxation time λ_1. Thus, the successful simulation of highly elastic flows and therefore, coping with the so called high Weissenberg number problem (HWNP) is a key focus of research in computational rheology [1].

The HWNP is attributed either to the limitations of the model or to errors in the numerical approximation scheme. Since for viscoelastic flow problems the existence and uniqueness cannot be proofed, it can happen to unknowingly simulate non-unique or inexistent solutions. Numerical errors can arise by choosing insufficient accurate numerical methods for the highly coupled, non-linear convection dominated and mixed hyperbolic-elliptic system of equations or by propagation of numerical oscillations if the discrete problem is not well-posed. This requires the satisfaction of compatibility conditions concerning the approximation spaces [1].

One good approach is the use of the Discontinuous Galerkin (DG) method having discontinuous elements with appropriate flux functions for the edges, which is more stable against numerical oscillations than e.g. Continuous Galerkin methods. Within the last 25 years, the DG method has been successfully established for solving hyperbolic conservation laws and was firstly introduced by Fortin and Fortin [2] for viscoelastic fluid flow. It is also strongly emerging in other fields of computational fluid dynamics [3]. There are two reasons for this ascent which obviates essential limitations of classical techniques such as finite volume or finite difference methods. DG cleverly combines

- an arbitrary order $k \in \mathbb{N}$ in the numerical discretization error $\mathcal{O}(h^k)$ with
- a local flux evaluation which is at most to be computed from adjacent cells.

Here h refers to the local grid spacing, and k to the order of the DG basis polynomials. This is in strong contrast to the established schemes, which are substantially limited to $\mathcal{O}(h^n)$ with $n \leq 2$ for unstructured grids, and even on Cartesian grids n is rather limited to small values because of the larger stencils for increasing n.

In viscoelastic flow the DG method is used to obtain stability for the convection dominated convection-diffusion problem using a streamline upwinding formulation. Therefore, the convective term of the constitutive equations is discretized by a DG scheme whereas the other terms as well as the momentum- and continuum equation are discretized using a standard finite element method (FEM). The DG method allows jumps in the boundary conditions and preconditioning at the elemental level, appropriate flux functions for the edges can be chosen and additional velocity-stress compatibility conditions can be easily satisfied [1]. So the DG method is a promising method for convection dominated problems. However, in case of viscoelastic flow there is no CFD-solver fully based on this method.

A breakdown in convergence can also occur due to the mixed hyperbolic-elliptic type of the system of equations. While the saddle point problem of the Navier-Stokes system is of elliptic type, the constitutive equations modelling the viscoelastic behaviour are hyperbolic. As it is shown in Sect. 2, the viscous part of the momentum equations and of the constitutive equations are weighted by a material parameter β. If β is close to 1, the contribution of the constitutive equations is small and we have an elliptical system to solve. If $\beta \to 0$ such that we have a highly viscoelastic fluid, a change of type from elliptic to hyperbolic occurs and the numerical solution becomes unstable [1].

There are several approaches in numerical computation for handling the strong mixed hyperbolic-elliptic coupling between the momentum and constitutive equations by the velocity gradient. In the elastic-viscous stress splitting (EVSS) method and its derivatives the depending variables are changed such that there is no necessity for additional compatibility conditions for the well-posedness of the discrete system in the Stokesian limit [1]. However, this extends the system of equations to be solved by an additional evolutionary equation for the velocity gradient.

We aim for a solver for viscoelastic flow with an exclusively high-order DG scheme for all equations using a local DG formulation with penalized fluxes in order to solve the hyperbolic constitutive equations and using a streamline upwinding for the convective fluxes of the constitutive equations.

The solver is embedded in the in-house DG framework called *BoSSS*, currently under development at the chair of fluid dynamics of TU Darmstadt.

2 Governing Equations

We consider two-dimensional incompressible viscoelastic Oldroyd B fluid flows in dimensionless variables consisting of the continuity equation

$$\nabla \cdot \mathbf{u} = 0, \tag{1}$$

the momentum equations including the solvent part of the fluid

$$\frac{\partial \mathbf{u}}{\partial t} + \mathbf{u} \cdot \nabla \mathbf{u} = -\nabla p + \frac{\beta}{\mathrm{Re}} \nabla \cdot \mathbf{D} + \frac{1}{\mathrm{Re}} \nabla \cdot \mathbf{T}, \tag{2}$$

and the constitutive equations for the polymeric part of the Oldroyd B fluid

$$\mathbf{T} + \mathrm{We}\, \overset{\nabla}{\mathbf{T}} = 2\,(1-\beta)\,\mathbf{D} \tag{3}$$

with Re: Reynolds number, We: Weissenberg number, β: ratio between retardation and relaxation time ($\beta = \frac{\lambda_2}{\lambda_1}$ with λ_1: relaxation time and λ_2: retardation time) and $\mathbf{D} = \frac{1}{2}(\nabla \mathbf{u} + (\nabla \mathbf{u})^T)$: rate of deformation tensor. We use the upper convected objective derivative:

$$\overset{\nabla}{\mathbf{T}} = \frac{\partial \mathbf{T}}{\partial t} + \mathbf{u} \cdot \nabla \mathbf{T} - \nabla \mathbf{u} \cdot \mathbf{T} - \mathbf{T} \cdot (\nabla \mathbf{u})^T . \tag{4}$$

For the splitting of the Oldroyd B model into a purely viscous solvent and a viscoelastic polymeric part and the non-dimensionalisation see e.g. [4]. The constitutive model reduces to the upper convected Maxwell model (Maxwell fluid B) if $\beta = 0$ and to a Newtonian fluid if $\beta = 1$ which means that the relaxation time and the retardation time are equal. If the Weissenberg number is chosen to be zero, this leads to a Newtonian formulation as well.

3 Discretization

At this point we want to focus on the discretization of the constitutive equations (3) and on the local DG formulation for the viscosity term on the right hand side of (3). For the viscosity term in the momentum part (2) the symmetric interior penalty method (SIP) is used [5]. For further information on the discretization of the continuity and the momentum equations read [5] and [6].

We choose $\beta = 0$ in Eqs. (2) and (3) and We $= 0$ in Eq. (3) which leads to Newtonian flow. Additionally, we restrict on Stokes flow for the momentum equations. For the discretization we consider six scalar equations with $(i, j) = 1, 2$ and $T_{12} = T_{21}$:

$$\begin{aligned}
\frac{\partial u_i}{\partial x_i} &= 0 && \text{in } \Omega \\
-\frac{\partial p}{\partial x_i} + \frac{1}{\text{Re}} \frac{\partial T_{ij}}{\partial x_j} &= 0 && \text{in } \Omega \\
T_{11} - 2\frac{\partial u_1}{\partial x_1} &= 0 && \text{in } \Omega \\
T_{12} - \frac{\partial u_1}{\partial x_2} - \frac{\partial u_2}{\partial x_1} &= 0 && \text{in } \Omega \\
T_{22} - 2\frac{\partial u_2}{\partial x_2} &= 0 && \text{in } \Omega \\
u_i &= u_{Di} && \text{on } \partial\Omega
\end{aligned} \tag{5}$$

We follow [7] to find the local DG formulation for this Newtonian fluid. We discretize the system (5) using DG. Therefore, we multiply all terms by arbitrary test functions v_l, $(l = 1, \ldots 6)$. After integrating by parts over a spatial element K in Ω we obtain the weak formulation for the continuity equation:

$$-\int_K u_i \frac{\partial v_1}{\partial x_i} \, dV + \int_{\partial K} u_i \, n_i \, v_1 \, dS = 0. \tag{6}$$

The weak formulation for the momentum part is:

$$\begin{aligned}
&\int_K p \frac{\partial v_2}{\partial x_1} \, dV - \int_{\partial K} p \, v_2 \, n_1 \, dS - \frac{1}{\mathrm{Re}} \int_K \left(T_{11} \frac{\partial v_2}{\partial x_1} + T_{12} \frac{\partial v_2}{\partial x_2} \right) dV \\
&\qquad - \frac{1}{\mathrm{Re}} \int_{\partial K} (T_{11} \, n_1 + T_{12} \, n_2) \, v_2 \, dS = 0 \\
&\int_K p \frac{\partial v_3}{\partial x_2} \, dV - \int_{\partial K} p \, v_3 \, n_2 \, dS - \frac{1}{\mathrm{Re}} \int_K \left(T_{12} \frac{\partial v_3}{\partial x_1} + T_{22} \frac{\partial v_3}{\partial x_2} \right) dV \\
&\qquad - \frac{1}{\mathrm{Re}} \int_{\partial K} (T_{12} \, n_1 + T_{22} \, n_2) \, v_3 \, dS = 0
\end{aligned} \tag{7}$$

and for the constitutive part:

$$\begin{aligned}
&\int_K T_{11} \, v_4 \, dV + 2 \int_K u_1 \frac{\partial v_4}{\partial x_1} \, dV + 2 \int_{\partial K} u_1 \, n_1 \, v_4 \, dS = 0 \\
&\int_K T_{12} \, v_5 \, dV + \int_K \left(u_1 \frac{\partial v_5}{\partial x_2} + u_2 \frac{\partial v_5}{\partial x_1} \right) dV \\
&\qquad + \int_{\partial K} (u_1 \, n_2 \, v_5 + u_2 \, n_1 \, v_5) \, dS = 0 \\
&\int_K T_{22} \, v_6 \, dV + 2 \int_K u_2 \frac{\partial v_6}{\partial x_2} \, dV + 2 \int_{\partial K} u_2 \, n_2 \, v_6 \, dS = 0
\end{aligned} \tag{8}$$

where $\mathbf{n}$ is the outward pointing unit vector normal to ∂K with its components n_i. For the approximate solution we have spatial discretized elements with $\mathcal{K}_h = \{K_1, \ldots K_j\}$ where K_j may not overlap and $\bigcup_j \bar{K}_j = \bar{\Omega}$. On this mesh we define the DG space $\mathbb{P}_k(\mathcal{K}_h)$ with broken polynomials of degree k. For the trial functions $(p_h, u_{hi}, T_{hij}) \in \mathbb{P}_k(\mathcal{K}_h)^2 \times \mathbb{P}_{k-1}(\mathcal{K}_h) \times \mathbb{P}_k(\mathcal{K}_h)^3$ we have to introduce numerical fluxes. These are for the continuity equation

$$-\sum_j \int_{K_j} u_{hi} \frac{\partial v_1}{\partial x_i} \, dV + \sum_j \int_{\partial K_j} \hat{f}(u_{hi}, n_i, v_1) \, dS = 0, \tag{9}$$

for the momentum equations

$$\begin{aligned}
&\sum_j \int_{K_j} p_h \frac{\partial v_2}{\partial x_1} \, dV - \sum_j \int_{\partial K_j} \hat{f}(p_h, n_1, v_2) \, dS \\
&-\sum_j \int_{K_j} \left(T_{h11} \frac{\partial v_2}{\partial x_1} + T_{h12} \frac{\partial v_2}{\partial x_2} \right) dV - \sum_j \int_{\partial K_j} \hat{f}(T_{h11}, T_{h12}, n_i, v_2) \, dS = 0 \\
&\sum_j \int_{K_j} p_h \frac{\partial v_3}{\partial x_2} \, dV - \sum_j \int_{\partial K_j} \hat{f}(p_h, n_2, v_3) \, dS \\
&-\sum_j \int_{K_j} \left(T_{h12} \frac{\partial v_3}{\partial x_1} + T_{h22} \frac{\partial v_3}{\partial x_2} \right) dV - \sum_j \int_{\partial K_j} \hat{f}(T_{h12}, T_{h22}, n_i, v_3) \, dS = 0,
\end{aligned} \tag{10}$$

11.5 (10)

and for the constitutive equations

$$\begin{aligned}
&\sum_j \int_{K_j} T_{h11} \, v_4 \, dV + \sum_j \int_{K_j} 2 \, u_{h1} \frac{\partial v_4}{\partial x_1} \, dV + \sum_j \int_{\partial K_j} 2 \, \hat{f}(u_{h1}, n_1, v_4) \, dS = 0 \\
&\sum_j \int_{K_j} T_{h12} \, v_5 \, dV \\
&+\sum_j \int_{K_j} \left(u_{h1} \frac{\partial v_5}{\partial x_2} + u_2 \frac{\partial v_5}{\partial x_1} \right) \, dV + \sum_j \int_{\partial K_j} \hat{f}(u_{hi}, n_i, v_5) \, dS = 0 \\
&\sum_j \int_{K_j} T_{h22} \, v_6 \, dV + \sum_j \int_{K_j} 2 \, u_{h2} \frac{\partial v_6}{\partial x_2} \, dV + \sum_j \int_{\partial K_j} 2 \, \hat{f}(u_{h2}, n_2, v_6) \, dS = 0.
\end{aligned} \tag{11}$$

We formulate the fluxes in a global edge-based manner such that the normal vector $\mathbf{n}$ is always pointing outward the considered cell (K^-) and into the neighbouring element (K^+). In advance, the mean values and jumps on the edges need to be defined:

$$\{a\} = \frac{1}{2}(a^- + a^+), \quad [\![a]\!] = a^- - a^+. \tag{12}$$

The fluxes are defined for an inner edge between K^- and K^+ as follows:

$$
\begin{aligned}
\hat{f}(u_{hi}, n_i, v_1) &= \alpha_2 [\![p]\!] [\![v_1]\!] - [\![u_i]\!] \cdot n_i \, \{v_1\} \\
\hat{f}(p_h, n_1, v_2) &= \{p\} \, n_1 [\![v_2]\!] \\
\hat{f}(p_h, n_2, v_3) &= \{p\} \, n_2 [\![v_3]\!] \\
\hat{f}(T_{h11}, T_{h12}, n_i, v_2) &= \left(\{T_{11}\} \, n_1 + \{T_{12}\} \, n_2 - \alpha_1 [\![u_1]\!]\right) [\![v_2]\!] \\
\hat{f}(T_{h12}, T_{h22}, n_i, v_3) &= \left(\{T_{12}\} \, n_1 + \{T_{22}\} \, n_2 - \alpha_1 [\![u_2]\!]\right) [\![v_3]\!] \\
\hat{f}(u_{h1}, n_1, v_4) &= 2 \, \{u_1\} \, n_1 [\![v_4]\!] \\
\hat{f}(u_{hi}, n_i, v_5) &= (\{u_1\} \, n_2 + \{u_2\} \, n_1) \, [\![v_5]\!] \\
\hat{f}(u_{h2}, n_2, v_6) &= 2 \, \{u_2\} \, n_2 [\![v_6]\!].
\end{aligned}
\tag{13}
$$

The fluxes for the boundary edges are defined for Dirichlet and Neumann boundary conditions on disjoint partitions of the boundary $\partial\Omega = \Gamma_D \bigcup \Gamma_N$:

$$
\begin{aligned}
\hat{f}(u_{hi}, n_i, v_1) &= 0 && \text{on } \Gamma_N \\
\hat{f}(u_{hi}, n_i, v_1) &= -(u_i^- - u_{Di}) \cdot n_i \, v_1^- && \text{on } \Gamma_D \\
\hat{f}(p_h, n_1, v_2) &= p^- \, n_1 v_2^- && \text{on } \Gamma_N \\
\hat{f}(p_h, n_1, v_2) &= p_D \, n_1 v_2^- && \text{on } \Gamma_D \\
\hat{f}(p_h, n_2, v_3) &= p^- \, n_2 v_3^- && \text{on } \Gamma_N \\
\hat{f}(p_h, n_2, v_3) &= p_D \, n_2 v_3^- && \text{on } \Gamma_D \\
\hat{f}(T_{h11}, T_{h12}, n_i, v_2) &= \left(T_{11}^- \, n_1 + T_{12}^- \, n_2\right) v_2^- && \text{on } \partial\Omega \\
\hat{f}(T_{h12}, T_{h22}, n_i, v_3) &= \left(T_{12}^- \, n_1 + T_{22}^- \, n_2\right) v_2^- && \text{on } \partial\Omega \\
\hat{f}(u_{h1}, n_1, v_4) &= 2 \, u_1^- \, n_1 v_4^- && \text{on } \Gamma_N \\
\hat{f}(u_{h1}, n_1, v_4) &= 2 \, u_{D1} \, n_1 v_4^- && \text{on } \Gamma_D \\
\hat{f}(u_{hi}, n_i, v_5) &= \left(u_1^- \, n_2 + u_2^- \, n_1\right) v_5^- && \text{on } \Gamma_N \\
\hat{f}(u_{hi}, n_i, v_5) &= (u_{D1} \, n_2 + u_{D2} \, n_1) \, v_5^- && \text{on } \Gamma_D \\
\hat{f}(u_{h2}, n_2, v_6) &= 2 \, u_2^- \, n_2 v_6^- && \text{on } \Gamma_N \\
\hat{f}(u_{h2}, n_2, v_6) &= 2 \, u_{D2} \, n_2 v_6^- && \text{on } \Gamma_D
\end{aligned}
\tag{14}
$$

The α_i are the penalties to ensure the stability of the system. They are chosen as follows:

$$\alpha_1 = \frac{s_1}{h_{\min}} \qquad \alpha_2 = s_2 \operatorname{Re} h_{\max}, \tag{15}$$

where $(h_{\min}, h_{\max})$ are the minimum and the maximum grid spacing. In this study the factors (s_1, s_2) are set to be 1, but they can enhance the penalization of the jumps between cells to improve stabilization of the system and to accelerate convergence.

For further information about the local DG formulation and about additional penalties with purpose of reducing sparsity and enhancing accuracy read [7].

4 Numerical Results

4.1 h^k Convergence Study

To proof convergence of the local DG formulation we use a manufactured non-polynomial divergence free solution introduced by [7]:

$$\begin{aligned} u_1(x_1, x_2) &= -e^{x_1} (x_2 \cos(x_2) + \sin(x_2)) \\ u_2(x_1, x_2) &= e^{x_1} x_2 \sin(x_2) \\ p(x_1, x_2) &= 2e^{x_1} \sin(x_2) \end{aligned} \tag{16}$$

The domain is $\Omega = (-1, 1)^2$ and the Reynolds number is chosen to be $\mathrm{Re} = 1$. In this case a source term on the right-hand side is not necessary. The mesh of the domain is uniform with quadratic elements of size $h = 2^{-\xi+1}$ where h is the grid spacing and ξ is the grid level. The manufactured solution (16) is non-polynomial such that it cannot be displayed exactly by the test functions v_l. For all boundaries a velocity inlet boundary condition is chosen.

The grid convergence is shown in Fig. 1 for different polynomial orders. They are in good agreement with the predicted order of convergence of $k + 1$ for the velocity and k for the pressure and the stresses, respectively, whereas k is the polynomial order of the test and trial functions.

In Table 1 our results are compared to the results of [7]. For first order polynomials as test functions we do not reach second order convergence for the velocity and stresses. For all other polynomial orders we reach higher convergence orders than predicted and than presented in [7].

4.2 Confined Cylinder

For testing Navier-Stokes flow a classical benchmark problem for viscoelastic fluid flow is presented: the confined cylinder. The two dimensional domain consist of a

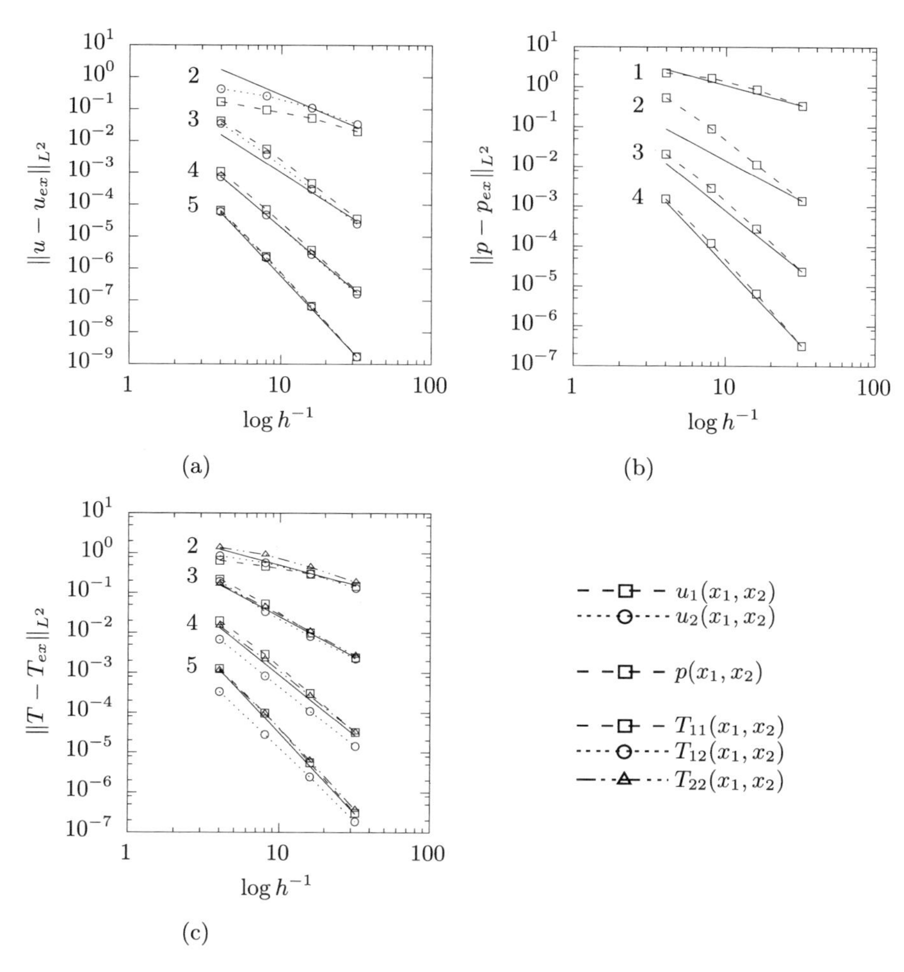

Fig. 1 h-convergence for the L_2-error against the analytical solution in Eq. (16). (**a**) Velocity. (**b**) Pressure. (**c**) Stresses

Table 1 Convergence order of Cockburn et al. [7] and of our LDG solver (here)

Degree	$\|u - u_{ex}\|_{L^2}$		$\|p - p_{ex}\|_{L^2}$		$\|T - T_{ex}\|_{L^2}$	
p	[7]	Here	[7]	Here	[7]	Here
1	2.04	1.18	1.36	0.91	0.82	0.90
2	3.05	3.44	2.18	2.88	1.85	2.11
3	4.06	4.10	2.77	3.27	2.80	3.04
4	–	5.05	–	4.09	–	3.92

wall bounded channel with height $a = 4$ and length $b = 30$ and a cylinder with a center position $(x_1, x_2) = (0, 0)$ and a radius $R = 0.25 \cdot a$, shown in Fig. 2. The grid has square elements which are adjusted to the boundary edges and gradually refined close to the cylinder. We consider steady Navier-Stokes flow such that we choose

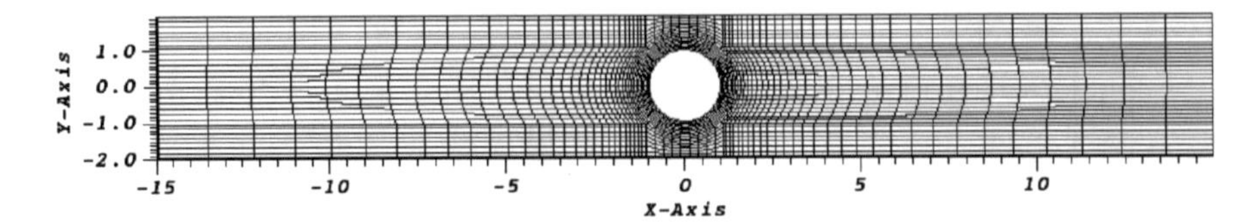

Fig. 2 Computational mesh of the confined cylinder benchmark problem with wall boundary conditions on the top and bottom, velocity inlet on the left and pressure outlet on the right. Channel height: $a = 4$, length: $b = 30$, cylinder center position: $(x_1, x_2) = (0, 0)$ and cylinder radius: $r = 1$

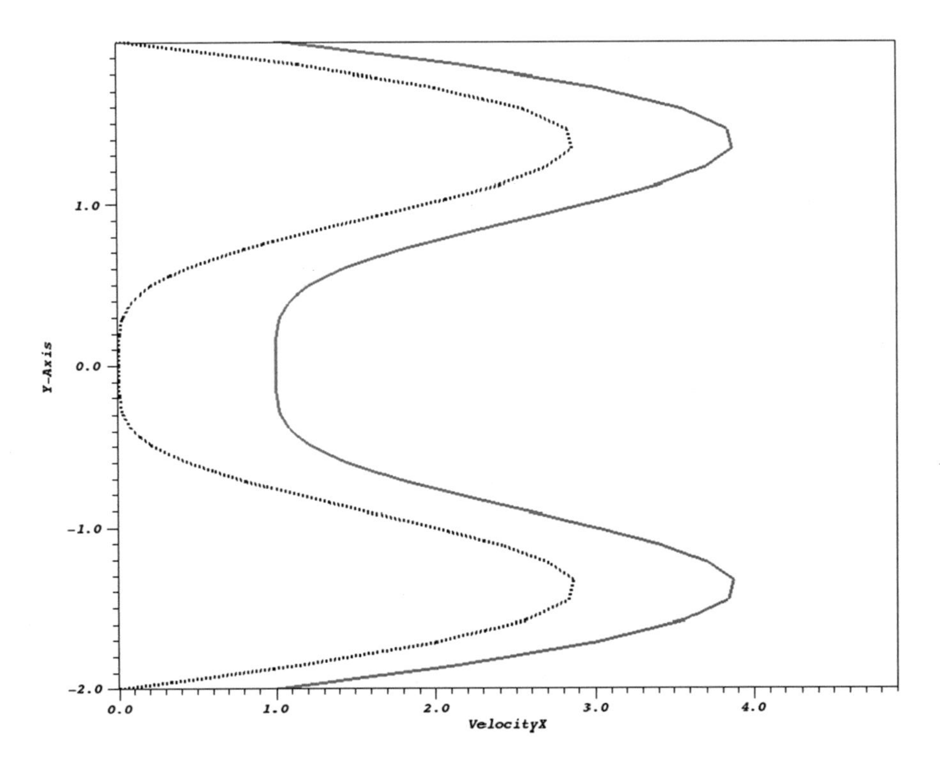

Fig. 3 Slice curve of $u_1(x_1, x_2)$ at cylinder position with SIP result in dotted black lines and LDG result in grey. The LDG result was shifted for better distinction between the lines

We $= 0$ for Newtonian flow. The boundary conditions are as follows: wall boundary conditions on the top and bottom, velocity inlet on the left and pressure outlet on the right. We can use β as a switching parameter between the SIP flux ($\beta = 1$) and the presented LDG formulation ($\beta = 0$).

As we can see in Fig. 3 the results hardly differ from each other. So even for sophisticated steady Navier-Stokes flow the LDG formulation shows reasonable results.

5 Concluding Remarks and Future Work

The purpose of our work is to develop a new numerical scheme for viscoelastic fluid flow exclusively based on DG methods. Two major problems concerning stability arise in simulating viscoelastic flow. First, we have a mixed hyperbolic-elliptic problem weighted by β and second, we have a convection-dominated convection-diffusion problem in the constitutive equations. We successfully implemented a local DG formulation with appropriate fluxes containing stabilizing penalty parameters following [7]. This stabilizes the solver in handling the possible change of type and calculating a hyperbolic problem.

Currently, the convective term of the constitutive equations is implemented in the solver. Therefore, we are using a streamline upwinding formulation for the convection-dominated problem following [1] and [6].

In a last step to a fully viscoelastic Oldroyd B fluid solver the objective term of the constitutive equations which behaves like a source term must be implemented and balanced together with the convective term.

Acknowledgements This work is partially supported by the 'Excellence Initiative' of the German Federal and State Governments within the Graduate School of Computational Engineering at Technische Universität Darmstadt.

References

1. Owens, R.G., Phillips, T.N.: Computational Rheology. Imperial College Press, London (2002)
2. Fortin, M., Fortin, A.: A new approach for the FEM simulation of viscoelastic flows. J. Non. Newton. Fluid. Mech. **32**(3), 295–310 (1989). https://doi.org/10.1016/0377-0257(89)85012-8
3. Cockburn, B., Karniadakis, G.E., Shu, C.-W.: Discontinuous Galerkin Methods. Lecture Notes in Computational Science and Engineering (11). Springer, Berlin (2000). https://doi.org/10.1007/978-3-642-59721-3
4. Phillips, T.N., Williams, A.J.: Viscoelastic flow through a planar contraction using a semi-Lagrangian finite volume method. J. Non. Newton. Fluid. Mech. **87**(2–3), 215–246 (1999). https://doi.org/10.1016/S0377-0257(99)00065-8
5. Di Pietro, D.A., Ern, A.: Mathematical Aspects of Discontinuous Galerkin Methods. Mathématiques et Applications, vol. 69. Springer, Berlin (2011). https://doi.org/10.1007/978-3-642-22980-0
6. Li, B.Q.: Discontinuous Finite Elements in Fluid Dynamics and Heat Transfer. Computational Fluid and Solid Mechanics. Springer, London (2006). https://doi.org/10.1007/1-84628-205-5
7. Cockburn, B., Kanschat, G., Schötzau, D., Schwab, C.: Local discontinuous galerkin methods for the stokes system. SIAM J. Numer. Anal. **40**(1), 319–343 (2002). https://doi.org/10.1137/S0036142900380121

A Bramble-Pasciak Conjugate Gradient Method for Discrete Stokes Problems with Lognormal Random Viscosity

Christopher Müller, Sebastian Ullmann, and Jens Lang

Abstract We study linear systems of equations arising from a stochastic Galerkin finite element discretization of saddle point problems with random data and its iterative solution. We consider the Stokes flow model with random viscosity described by the exponential of a correlated random process and shortly discuss the discretization framework and the representation of the emerging matrix equation. Due to the high dimensionality and the coupling of the associated symmetric, indefinite, linear system, we resort to iterative solvers and problem-specific preconditioners. As a standard iterative solver for this problem class, we consider the block diagonal preconditioned MINRES method and further introduce the Bramble-Pasciak conjugate gradient method as a promising alternative. This special conjugate gradient method is formulated in a non-standard inner product with a block triangular preconditioner. From a structural point of view, such a block triangular preconditioner enables a better approximation of the original problem than the block diagonal one. We derive eigenvalue estimates to assess the convergence behavior of the two solvers with respect to relevant physical and numerical parameters and verify our findings by the help of a numerical test case. We model Stokes flow in a cavity driven by a moving lid and describe the viscosity by the exponential of a truncated Karhunen-Loève expansion. Regarding iteration numbers, the Bramble-Pasciak conjugate gradient method with block triangular preconditioner is superior to the MINRES method with block diagonal preconditioner in the considered example.

C. Müller (✉) · S. Ullmann · J. Lang
Graduate School of Computational Engineering, Technische Universität Darmstadt, Darmstadt, Germany

Department of Mathematics, Technische Universität Darmstadt, Darmstadt, Germany
e-mail: cmueller@gsc.tu-darmstadt.de; http://www.graduate-school-ce.de/index.php?id=680

M. Schäfer et al. (eds.), *Recent Advances in Computational Engineering*, Lecture Notes in Computational Science and Engineering 124,
https://doi.org/10.1007/978-3-319-93891-2_5

Keywords Uncertainty quantification · PDEs with random data · Stokes flow · Preconditioning · Stochastic Galerkin · Lognormal data · Mixed finite elements · Conjugate gradient method · Saddle point problems

1 Introduction

We study the stochastic Galerkin finite element (SGFE) method [2, 13, 14] as a tool to approximate statistical quantities in the context of saddle point problems with random input data. Stochastic Galerkin (SG) methods rely on a representation of the random input based on a vector of random variables with known probability density. The starting point of the method is a weak formulation not only over the spatial domain but also over the image domain of this random vector. The SGFE approach enables the computation of the unknown solution coefficients via a Galerkin projection onto the finite-dimensional tensor product space of a finite element (FE) space for the spatial dependencies and a global polynomial/SG space for the random vector.

Concerning convergence rates, SG methods are often superior to more robust stochastic methods—such as Monte-Carlo sampling—when the input data exhibits certain regularity structures [4, 16]. This is a property shared by the main competitors of the SG approaches: Stochastic collocation methods [3, 14] similarly exploit the structure of the random input and further rely on uncoupled solutions of the underlying deterministic problem, just like sampling methods. An advantage of SG methods is that rigorous error analysis can be used to analyze them, but they are more challenging from a computational point of view: A block structured system of coupled deterministic problems must be solved. However, this can be done efficiently using iterative methods and problem-specific preconditioners, see e.g. [11, 19, 22, 24, 29].

We build on these results and consider the Bramble-Pasciak conjugate gradient (BPCG) method [7] in the SGFE setting with lognormal data. We compare it to the MINRES approach, the standard Krylov subspace solver for the problem class we consider, and investigate the performance of both solvers with respect to different problem and discretization parameters. This is done both analytically and based on a numerical test case.

As a saddle point problem, we consider the Stokes flow model in a bounded domain $D \subset \mathbb{R}^2$ with boundary ∂D. The vector $x = (x_1, x_2)^{\mathrm{T}} \in D$ denotes the spatial coordinates. The Stokes equations are a simplification of the Navier-Stokes equations and describe the behavior of a vector-valued velocity field $u = (u_1, u_2)^{\mathrm{T}}$ and a scalar pressure field p subject to viscous and external forcing. We also introduce the probability space $(\Omega, \mathcal{F}, \mathbb{P})$, where Ω denotes the set of elementary events, $\mathcal{F}$ is a σ-algebra on Ω and $\mathbb{P} : \mathcal{F} \to [0, 1]$ is a probability measure.

Input data are inherently uncertain due to either a lack of knowledge or simply imprecise measurements. Taking into account this variability in the model, we assume that the viscosity is a random field $\nu = \nu(x, \omega)$, $\nu : D \times \Omega \to \mathbb{R}$. Since the input uncertainty propagates through the model, the solution components also have to be considered as random fields. In summary, the strong form of the Stokes equations with uncertain viscosity is the following:

Find $u = u(x, \omega)$ and $p = p(x, \omega)$ such that, $\mathbb{P}$-almost surely,

$$\begin{aligned} -\nabla \cdot \big(\nu(x, \omega)\, \nabla u(x, \omega)\big) + \nabla p(x, \omega) &= f(x) \quad \text{in } D \times \Omega, \\ \nabla \cdot u(x, \omega) &= 0 \qquad \text{in } D \times \Omega, \\ u(x, \omega) &= g(x) \quad \text{on } \partial D \times \Omega. \end{aligned} \tag{1}$$

Both, the volume force $f = (f_1, f_2)^{\mathrm{T}}$ and the boundary data $g = (g_1, g_2)^{\mathrm{T}}$ are assumed to be deterministic functions. This is for the sake of simplicity of notation. Treating stochastic forcing and boundary data in the model would be straightforward under appropriate integrability assumptions on the data.

The remainder of the paper is organized as follows: As a model for the uncertain viscosity, we introduce the exponential of a Karhunen-Loève expansion (KLE) of a Gaussian random field in Sect. 2. We ensure boundedness of this expansion by stating suitable assumptions on its components. The boundedness of the viscosity is necessary for the well-posedness of the variational formulation we introduce in Sect. 3. In Sect. 4, we establish a matrix representation of the Stokes problem with lognormal random data by restricting the weak equations to a finite-dimensional subspace spanned by Taylor-Hood finite elements and Hermite chaos polynomials. Preconditioning strategies are discussed in Sect. 5 where we consider block diagonal and block triangular preconditioning structures. As building blocks, we use a Kronecker product structure with established approaches from the FE and SG literature as input. In Sect. 6, we derive inclusion bounds for the eigenvalues of relevant sub-matrices and interpret them concerning the overall convergence behavior. By modifying our block triangular preconditioner in a specific way, we ensure the existence of a conjugate gradient (CG) method in a non-standard inner product. We discuss the application of this CG method as well as the application of the MINRES iterative solver in greater detail in Sect. 7. A numerical test case is considered in Sect. 8 where we illustrate the expected convergence behavior of the two considered solvers with respect to different problem parameters. The final section eventually summarizes and concludes our work.

2 Input Modeling

We start our considerations with the random field $\mu(x, \omega)$ and assume that it is Gaussian and second-order, meaning $\mu \in L^2(\Omega, L^2(D))$. In this setting it is possible to represent $\mu(x, \omega)$ as a KLE of the form [18, Theorem 5.28]

$$\mu(x, \omega) = \mu_0(x) + \sigma_\mu \sum_{m=1}^{\infty} \sqrt{\lambda_m} \mu_m(x) y_m(\omega). \tag{2}$$

Here, $\mu_0(x)$ is the mean field of $\mu(x, \omega)$, i.e. $\mu_0(x) = \int_\Omega \mu(x, \omega)\, \mathrm{d}\mathbb{P}(\omega)$ is the expected value of the random field. The eigenpairs $(\lambda_m, \mu_m)_{m=1}^{\infty}$ belong to the integral operator which corresponds to the covariance function of the correlated Gaussian random field. Further, $y_m, m = 1, \dots, \infty$, are uncorrelated Gaussian random variables with zero mean and unit variance which live in the unbounded set of sequences $\mathbb{R}^{\mathbb{N}}$. As we are in the Gaussian setting, the random variables originating from the KLE are also stochastically independent.

The unbounded support of the Gaussian random variables leads to one major problem concerning the theoretical investigations of the problem: As there is always a nonzero probability that random variables take on negative values with arbitrarily large magnitudes, negative values of the modeled viscosity can occur independent of its construction. We use a standard approach to avoid negative values and apply the exponential function to (2), yielding

$$\nu(x, \omega) = \exp(\mu(x, \omega)) = \exp\left(\mu_0(x) + \sigma_\mu \sum_{m=1}^{\infty} \sqrt{\lambda_m} \mu_m(x) y_m(\omega)\right), \tag{3}$$

for $y = (y_m(\omega))_{m \in \mathbb{N}}$, $y : \Omega \to \mathbb{R}^{\mathbb{N}}$. Expression (3) is called a lognormal random process as the logarithm of $\nu(x, \omega)$ is the Gaussian process $\mu(x, \omega)$. In order to ensure boundedness of the viscosity, some assumptions have to be made on the series components. Following [16], we assume that

1. the mean field of $\mu(x, \omega)$ and the product of the Karhunen-Loève eigenpairs is bounded:

$$\mu_0, \sqrt{\lambda_m} \mu_m \in L^\infty(D), \quad \forall m \in \mathbb{N}, \tag{4}$$

2. the series of the product of the Karhunen-Loève eigenpairs converges absolutely:

$$\chi := (\chi_m)_{m \geq 1} = \left(\|\sqrt{\lambda_m} \mu_m(x)\|_{L^\infty(D)}\right)_{m \geq 1} \in \ell^1(\mathbb{N}). \tag{5}$$

Then, we define the set

$$\Xi^M := \mathbb{R}^M.$$

The set Ξ^M is M-dimensional but can also be defined for infinitely many parameters. However, we will not do this here as it introduces additional difficulties to the problem. For a full analysis, we refer to [25].

We define the value $\overline{\mu}_0 := \|\mu_0(x)\|_{L^\infty(D)}$ and the truncated viscosity:

$$\nu_M(x, y) = \exp\Big(\mu_0(x) + \sigma_\mu \sum_{m=1}^{M} \sqrt{\lambda_m}\mu_m(x)\, y_m\Big), \tag{6}$$

for $y \in \Xi^M$: Given the assumptions (4) and (5), the viscosity (6) satisfies

$$0 < \underline{\nu}(y) := \operatorname*{ess\,inf}_{x \in D} \nu_M(x, y) \leq \nu_M(x, y) \leq \operatorname*{ess\,sup}_{x \in D} \nu_M(x, y) =: \overline{\nu}(y), \tag{7}$$

with

$$\overline{\nu}(y) \leq \exp(\overline{\mu}_0) \exp\Big(\sum_{m=1}^{M} \chi_m |y_m|\Big),$$

$$\underline{\nu}(y) \geq \exp(-\overline{\mu}_0) \exp\Big(-\sum_{m=1}^{M} \chi_m |y_m|\Big),$$

for $y \in \Xi^M$. A proof of (7) can be found in [16, Lemma 2.2].

As a consequence of (7), the viscosity (3) is bounded from above and has a positive lower bound for almost all $\omega \in \Omega$. This is a basic property necessary for our problem to be well-defined. A reasonable truncation is possible with moderate index M when the covariance operator of the underlying random field is sufficiently smooth and the correlation length is sufficiently large.

The stochastic independence of the random variables allows us to express the corresponding Gaussian density as a product of univariate densities:

$$\rho(y) := \prod_{m=1}^{M} \rho_m(y_m), \quad \text{with} \quad \rho_m(y_m) := \frac{1}{\sqrt{2\pi}} \exp(-y_m^2/2). \tag{8}$$

It makes sense to change from the abstract domain Ω to the domain Ξ^M in the following, see [16, section 2.1]. The change of spaces leads to an integral transform, such that the computation of the pth moment of a random variable $v(\omega)$ is done via

$$\langle v^p \rangle := \int_{\Xi^M} v^p(y)\, \rho(y)\, \mathrm{d}y. \tag{9}$$

As a consequence of the Doob-Dynkin lemma [20], the output random fields can be parametrized with the vector y as well.

3 Variational Formulation

In order to formulate the weak equations derived from (1), we introduce Bochner spaces $L^2_\rho(\Xi^M; X)$, where X is a separable Hilbert space. They consist of all equivalence classes of strongly measurable functions $v : \Xi^M \to X$ with norm

$$\|v\|_{L^2_\rho(\Xi^M;X)} = \left(\int_{\Xi^M} \|v(\cdot, y)\|_X^2 \,\rho(y)\,\mathrm{d}y\right)^{1/2} < \infty.$$

In the following, we work in the tensor product spaces $L^2_\rho(\Xi^M) \otimes X$ with corresponding norms $\|\cdot\|_{L^2_\rho(\Xi^M)\otimes X} := \|\cdot\|_{L^2_\rho(\Xi^M;X)}$ as they are isomorphic to the Bochner spaces for separable X.

For the Stokes problem with random data, we insert the standard function spaces for enclosed flow and introduce

$$\begin{aligned} \mathcal{V}_0 &:= L^2_\rho(\Xi^M) \otimes \boldsymbol{V}_0(D), \\ \mathcal{W}_0 &:= L^2_\rho(\Xi^M) \otimes W_0(D), \end{aligned} \tag{10}$$

with

$$\begin{aligned} \boldsymbol{V}_0 &:= \boldsymbol{H}^1_0(D) = \left\{v \in \boldsymbol{H}^1(D) \mid v_{|\partial D} = 0\right\}, \\ W_0 &:= L^2_0(D) = \left\{q \in L^2(D) \mid \textstyle\int_D q(x)\,\mathrm{d}x = 0\right\}. \end{aligned}$$

The product spaces are Hilbert spaces as well.

We are now able to formulate the variational formulation associated with (1): Find $(u, p) \in \mathcal{V}_0 \times \mathcal{W}_0$ satisfying

$$\begin{aligned} \langle a(u, v)\rangle + \langle b(v, p)\rangle &= \langle l(v)\rangle, &\quad \forall v \in \mathcal{V}_0, \\ \langle b(u, q)\rangle &= \langle t(q)\rangle, &\quad \forall q \in \mathcal{W}_0, \end{aligned} \tag{11}$$

with bilinear forms

$$\langle a(u, v)\rangle := \int_{\Xi^M} \int_D \nu_M(x, y)\, \nabla u(x, y) \cdot \nabla v(x, y)\, \rho(y)\,\mathrm{d}x\,\mathrm{d}y, \tag{12}$$

$$\langle b(v, q)\rangle := -\int_{\Xi^M} \int_D q(x, y)\, \nabla \cdot v(x, y)\, \rho(y)\,\mathrm{d}x\,\mathrm{d}y, \tag{13}$$

for $u, v \in \mathcal{V}_0$, $q \in \mathcal{W}_0$, and linear functionals

$$\begin{aligned} \langle l(v)\rangle &= \int_{\Xi^M}\int_D f(x)\cdot v(x,y)\,\rho(y)\,\mathrm{d}x\,\mathrm{d}y - \langle a(u_0,v)\rangle, && v\in\mathcal{V}_0,\\ \langle t(v)\rangle &= -\langle b(u_0,q)\rangle, && q\in\mathcal{W}_0, \end{aligned}$$

where $u_0 \in \mathcal{V}_0$ is the lifting of the boundary data g in the sense of the trace theorem.

The weak Eq. (11) are a set of parametric deterministic equations which contain the full stochastic information of the original problem with random data. For the well-posedness of the variational formulation (11), we refer to [25].

4 Stochastic Galerkin Finite Element Discretization

We derive a discrete set of equations from (11) by choosing appropriate subspaces for the building blocks of the product spaces (10). For the discretization of the physical space, we use FE subspaces $V_0^h \subset V_0$ and $W_0^h \subset W_0$, where h denotes the mesh size. The domain of the random variables is discretized with generalized polynomial chaos [31]. The corresponding SG space is denoted by $S^k \subset L_\rho^2(\Xi^M)$, where k is the degree of the chaos functions and M is the truncation index of the KLE in (6). The SGFE subspaces $\mathcal{V}_0^{kh} \subset \mathcal{V}_0$ and $\mathcal{W}_0^{kh} \subset \mathcal{W}_0$ are now defined as products of the separate parts:

$$\begin{aligned} \mathcal{V}_0^{kh} &:= S^k \otimes V_0^h,\\ \mathcal{W}_0^{kh} &:= S^k \otimes W_0^h. \end{aligned} \tag{14}$$

As a choice for the spatial discretization, we use inf-sup stable Taylor-Hood P_2/P_1 finite elements on a regular triangulation. They consist of N_u continuous piecewise quadratic basis functions for the velocity space and N_p continuous piecewise linear basis functions for the pressure space. For the parametric space, we choose a discretization based on a complete multivariate Hermite polynomial chaos. The corresponding basis functions are global polynomials which are orthonormal with respect to the joint density $\rho(y)$ in (8). Therefore, they are the appropriate match to the Gaussian distribution of the input parameters according to the Wiener-Askey scheme [31]. We construct the M-variate basis functions as a product of M univariate chaos polynomials. We work with a complete polynomial basis, i.e. we choose a total degree k and the sum of the degrees of the M univariate chaos polynomials $\sum_{m=1}^M k_m$ must be less then or equal k. This yields a multivariate chaos or stochastic Galerkin basis of size Q_z, where

$$Q_z := \binom{M+k}{k}. \tag{15}$$

Behind every index $q = 0, \ldots, Q_z - 1$ there is a unique combination of univariate polynomial degrees—a multi-index—$(k_1, \ldots, k_M)$ and vice versa. The qth multivariate polynomial chaos basis function is thus the product of M univariate chaos polynomials with degrees from the qth multi-index.

Regarding the representation (6), a separation of the spatial and parametric dependencies would be beneficial. Such a decomposition can be achieved by projecting the exponential of the truncated KLE onto a Hermite chaos basis [24]:

$$\nu_M(x, y) = \sum_{q=0}^{Q_\nu - 1} \nu_q(x) \psi_q(y), \tag{16}$$

where $\{\psi_q(y)\}_{q=0}^{Q_\nu-1}$ are the Hermite chaos basis functions and

$$\nu_q(x) = \exp\left(\mu_0(x) + \tfrac{1}{2}\sigma_\mu^2 \sum_{m=1}^{M} \lambda_m \mu_m^2(x)\right) \prod_{m=1}^{M} \frac{\left(\sigma_\mu \sqrt{\lambda_m} \mu_m(x)\right)^{k_m}}{\sqrt{k_m!}}. \tag{17}$$

Again, there is a unique multi-index $(k_1, \ldots, k_M)$ to every index $q = 1, \ldots, Q_\nu$.

When used in a stochastic Galerkin setting, the representation (16) is in fact exact if we use the same Hermite polynomial chaos basis as for the representation of the solution fields but with twice the total degree [26, Remark 2.3.4], i.e.

$$Q_\nu := \binom{M + 2k}{2k}. \tag{18}$$

Although Q_ν grows fast with k and M and can be a lot bigger than Q_z, it does not make sense to truncate the sum in (16) prematurely. Doing so may destroy the coercivity of (12) and the corresponding discrete operator can easily become indefinite, see [26, Example 2.3.6]. Consequently, we always use all terms in (16).

Without going into the details, we will assume in the following that the fully discrete problem is well-posed which implies that the discretizations we choose are inf-sup stable on the discrete product spaces (14). An analysis of discrete inf-sup stability for a mixed formulation of the diffusion problem with uniform random data can be found in [6, Lemma 3.1].

The size of the emerging system of equations for our chosen discretizations is

$$\dim(\boldsymbol{\mathcal{V}}_0^{kh} \times \mathcal{W}_0^{kh}) = Q_z\,(N_u + N_p) = Q_z\,N. \tag{19}$$

To derive a matrix equation of the Stokes problem with random data, the velocity and pressure random fields as well as the test functions are represented in the FE and SG bases and subsequently inserted into the weak formulation (11) together with the input representation (16), yielding

$$\mathcal{C}\,\boldsymbol{w} = \boldsymbol{b}, \quad \mathcal{C} \in \mathbb{R}^{Q_z N \times Q_z N}, \tag{20}$$

where

$$\mathcal{C} := \begin{bmatrix} \mathcal{A} & \mathcal{B}^{\mathrm{T}} \\ \mathcal{B} & 0 \end{bmatrix}, \quad \boldsymbol{w} := \begin{bmatrix} \boldsymbol{u} \\ \boldsymbol{p} \end{bmatrix}, \quad \boldsymbol{b} := \begin{bmatrix} \boldsymbol{f} \\ \boldsymbol{t} \end{bmatrix}. \tag{21}$$

Here, the vectors $\boldsymbol{u} \in \mathbb{R}^{Q_z N_u}$ and $\boldsymbol{p} \in \mathbb{R}^{Q_z N_p}$ contain the coefficients of the discrete velocity and pressure solutions, respectively. Furthermore,

$$\mathcal{A} = I \otimes A_0 + \sum_{q=1}^{Q_\nu - 1} G_q \otimes A_q \qquad \in \mathbb{R}^{Q_z N_u \times Q_z N_u}, \tag{22}$$

$$\mathcal{B} = I \otimes B \qquad \in \mathbb{R}^{Q_z N_p \times Q_z N_u}, \tag{23}$$

$$\boldsymbol{f} = \boldsymbol{g}_0 \otimes \boldsymbol{w} \qquad \in \mathbb{R}^{Q_z N_u}, \tag{24}$$

$$\boldsymbol{t} = \boldsymbol{g}_0 \otimes \boldsymbol{d} \qquad \in \mathbb{R}^{Q_z N_p}, \tag{25}$$

where $I \in \mathbb{R}^{Q_z \times Q_z}$ is the Gramian of S^k, because the basis is orthonormal. Further, the $G_q \in \mathbb{R}^{Q_z \times Q_z}$ emerge from the evaluation of the product of three Hermite chaos basis functions in the expectation (9) with $p = 1$. They are called stochastic Galerkin matrices in the following. We call the $A_q \in \mathbb{R}^{N_u \times N_u}$ weighted FE velocity Laplacians as they are FE velocity Laplacians weighted with the functions $\nu_q(x)$, $q = 0 \ldots, Q_\nu - 1$. The matrices and vectors in (22)–(25) can all be constructed when the FE and SG basis representations are inserted into the variational formulation (11) together with the input representation (16). To avoid confusion, the matrices on the product spaces are calligraphic capital letters whereas the matrices on either the FE or the SG spaces are standard capital letters.

The size of the SGFE system is the product of the size of the FE basis N and the size of the SG basis Q_z. Additionally, the problem is coupled, as the symmetric matrices G_q, $q = 1 \ldots, Q_\nu - 1$, are not diagonal. Therefore, realistic problems are often too big to be treated with direct solution methods, which is why iterative algorithms are used instead. The application of iterative methods naturally raises the question of efficient preconditioning due to the inherent ill-conditioning of the problem. For the mentioned reasons, preconditioners and iterative methods are investigated in the following.

5 Preconditioning

The SGFE matrix $\mathcal{C} \in \mathbb{R}^{Q_z N \times Q_z N}$ in (21) is a symmetric saddle point matrix. Therefore, the following factorizations exist:

$$\begin{bmatrix} \mathcal{A} & \mathcal{B}^{\mathrm{T}} \\ \mathcal{B} & 0 \end{bmatrix} = \begin{bmatrix} I & 0 \\ \mathcal{B}\mathcal{A}^{-1} & I \end{bmatrix} \begin{bmatrix} \mathcal{A} & 0 \\ 0 & \mathcal{S} \end{bmatrix} \begin{bmatrix} I & \mathcal{A}^{-1}\mathcal{B}^{\mathrm{T}} \\ 0 & I \end{bmatrix} = \begin{bmatrix} \mathcal{A} & 0 \\ \mathcal{B} & \mathcal{S} \end{bmatrix} \begin{bmatrix} I & \mathcal{A}^{-1}\mathcal{B}^{\mathrm{T}} \\ 0 & I \end{bmatrix}, \tag{26}$$

where $\mathcal{S} := -\mathcal{B}\mathcal{A}^{-1}\mathcal{B}^{\mathrm{T}}$ is the SGFE Schur complement. In Sect. 4, we have assumed well-posedness of the discrete variational problem. This implies that the discrete version of the bilinear form (12) is continuous and coercive. Inserting the FE and SG basis, the coercivity condition translates into positive definiteness of the matrix $\mathcal{A}$. The congruence transform in (26) then implies that the discrete SGFE problem is indefinite as the Schur complement is negative semi-definite by construction.

Motivated by the factorizations (26), we consider block diagonal and block triangular preconditioning structures in the following. In the context of solving saddle point problems, these are well established concepts [5], which are generically given by

$$\mathcal{P}_1 = \begin{bmatrix} \widetilde{\mathcal{A}} & 0 \\ 0 & \widetilde{\mathcal{S}} \end{bmatrix}, \quad \mathcal{P}_2 = \begin{bmatrix} \widetilde{\mathcal{A}} & 0 \\ \mathcal{B} & \widetilde{\mathcal{S}} \end{bmatrix}. \tag{27}$$

Here, $\widetilde{\mathcal{A}}$ and $\widetilde{\mathcal{S}}$ are approximations of $\mathcal{A}$ and $\mathcal{S}$, respectively. Choosing these approximations appropriately is the main task of preconditioning. Desirable properties include a reduction of computational complexity and an improvement of the condition of the involved operators.

In the following, we do not directly look for suitable $\widetilde{\mathcal{A}}$ and $\widetilde{\mathcal{S}}$, but make another structural assumption prior to that. We want each SGFE preconditioner to be the Kronecker product of one FE and one SG preconditioner, i.e.

$$\widetilde{\mathcal{A}} := \widetilde{G}_A \otimes \widetilde{A}, \quad \widetilde{\mathcal{S}} := \widetilde{G}_S \otimes \widetilde{S}, \tag{28}$$

where $\widetilde{A} \in \mathbb{R}^{N_u \times N_u}$ and $\widetilde{S} \in \mathbb{R}^{N_p \times N_p}$ are approximations of the FE operators and the matrices $\widetilde{G}_A, \widetilde{G}_S \in \mathbb{R}^{Q_z \times Q_z}$ are approximations of the SG operators. The structural simplifications (28) have two advantages: Firstly, the Kronecker product is trivially invertible. Secondly, if we look into $\widetilde{\mathcal{A}}^{-1}\mathcal{A}$, we get

$$\widetilde{\mathcal{A}}^{-1}\mathcal{A} = \left(\widetilde{G}_A^{-1} \otimes \widetilde{A}^{-1} A_0 + \sum_{q=1}^{Q_\nu - 1} \widetilde{G}_A^{-1} G_q \otimes \widetilde{A}^{-1} A_q \right). \tag{29}$$

The SG preconditioner $\widetilde{G}_A$ only acts on the SG matrices I and G_q and the FE preconditioner $\widetilde{A}$ acts on the weighted FE Laplacians A_q, $q = 0, \dots, Q_\nu - 1$. This works in the same way for the preconditioned SGFE Schur complement, as we will see in Lemma 3 on page 78. The separation into FE and SG parts thus allows the use of established preconditioners from the FE and SG literature as building blocks for the SGFE preconditioners. Now, we will choose suitable approximations $\widetilde{A}$ and $\widetilde{S}$ to the FE Laplacian and Schur complement, respectively, and suitable approximations $\widetilde{G}_A$ and $\widetilde{G}_S$ to the SG matrices.

5.1 Finite Element Matrices

First of all, we consider a preconditioner for the FE Laplacian and derive bounds for the eigenvalues of the preconditioned weighted FE Laplacians. Then, we decide on a preconditioner for the FE Schur complement.

To precondition FE Laplacians, the multigrid method has emerged as one of the most suitable approaches. The first reason for that is the following: One multigrid V-cycle with appropriate smoothing—denoted by $\widetilde{A}_{\mathrm{mg}}$ in the following—is spectrally equivalent to the FE Laplacian A, see [8, section 2.5]. In this context, spectral equivalence means that there exist positive constants δ and Δ, independent of h, such that

$$\delta \le \frac{\boldsymbol{v}^{\mathrm{T}} A\, \boldsymbol{v}}{\boldsymbol{v}^{\mathrm{T}} \widetilde{A}_{\mathrm{mg}} \boldsymbol{v}} \le \Delta, \quad \forall \boldsymbol{v} \in \mathbb{R}^{N_u} \backslash \{\mathbf{0}\}. \tag{30}$$

As the eigenvalues of A actually depend on the mesh width h [8, Theorem 3.21], preconditioning with $\widetilde{A}_{\mathrm{mg}}$ thus eliminates this h-ill-conditioning of A. The multigrid method is also attractive from a computational point of view as it can be applied with linear complexity. Concerning (30), when the multigrid operator is applied to the weighted FE Laplacians in (22), we derive

$$-\,\overline{\nu}_q\, \Delta \le \frac{\boldsymbol{v}^{\mathrm{T}} A_q\, \boldsymbol{v}}{\boldsymbol{v}^{\mathrm{T}} \widetilde{A}_{\mathrm{mg}} \boldsymbol{v}} \le \overline{\nu}_q\, \Delta, \quad \forall \boldsymbol{v} \in \mathbb{R}^{N_u} \backslash \{\mathbf{0}\}, \tag{31}$$

where $\overline{\nu}_q := \|\nu_q(x)\|_{L^\infty(D)}$, for $q = 1, \ldots, Q_\nu - 1$. For $q = 0$, the lower bound is different, namely $\underline{\nu}_0 := \inf_{x \in D} \nu_0(x) > 0$. This bound is tighter because we know that the function $\nu_0 = \exp(\mu_0(x) + \frac{1}{2}\sigma_\mu^2 \sum_{m=1}^{M} \lambda_m \mu_m^2(x))$ from (17) is always positive. The functions are in $L^\infty(D)$ due to the assumptions (4) and the continuity of the exponential function.

The pressure mass matrix M_p is a good preconditioner for the negative FE Schur complement $-S = BA^{-1}B^{\mathrm{T}}$, because the matrices are spectrally equivalent in the sense that [8, Theorem 3.22]

$$\gamma^2 \le \frac{\boldsymbol{q}^{\mathrm{T}} B A^{-1} B^{\mathrm{T}} \boldsymbol{q}}{\boldsymbol{q}^{\mathrm{T}} M_p\, \boldsymbol{q}} \le 1, \quad \forall \boldsymbol{q} \in \mathbb{R}^{N_p} \backslash \{\mathbf{0}, \mathbf{1}\}, \tag{32}$$

where B is the FE divergence matrix and $\gamma > 0$ is the inf-sup constant of our mixed FE approximation. Further, the notation $\backslash\{\mathbf{0}, \mathbf{1}\}$ means we exclude all multiples of the constant function, see [8, Section 3.3]. If necessary, we always exclude the hydrostatic pressure in this way in the following. As M_p has the usual FE sparsity, using it as a preconditioner is too expensive in practice. Therefore, another approximation is considered, namely its diagonal $D_p := \mathrm{diag}(M_p)$. It is spectrally equivalent to M_p, i.e. there exist $\theta, \Theta > 0$ such that

$$\theta \leq \frac{\boldsymbol{q}^{\mathrm{T}} M_p \boldsymbol{q}}{\boldsymbol{q}^{\mathrm{T}} D_p \boldsymbol{q}} \leq \Theta, \quad \forall \boldsymbol{q} \in \mathbb{R}^{N_p} \backslash \{\mathbf{0}\}. \tag{33}$$

The constants θ and Θ only depend on the degree and type of finite elements used [30]. We use piecewise linear basis functions on triangles in our work, yielding $\theta = \frac{1}{2}$ and $\Theta = 2$. Consequently, D_p is spectrally equivalent to the negative FE Schur complement. This directly follows from (32) and (33):

$$\theta \gamma^2 \leq \frac{\boldsymbol{q}^{\mathrm{T}} B A^{-1} B^{\mathrm{T}} \boldsymbol{q}}{\boldsymbol{q}^{\mathrm{T}} D_p \boldsymbol{q}} \leq \Theta, \quad \forall \boldsymbol{q} \in \mathbb{R}^{N_p} \backslash \{\mathbf{0}, \mathbf{1}\}. \tag{34}$$

Using D_p as a preconditioner is also attractive from a complexity point of view as it can be applied with linear costs. The bounds (30), (31) and (34), which are independent of the mesh size, are the basis for analyzing the h-independence of the preconditioned SGFE saddle point problem later on in Sect. 6.

Due to the mentioned reasons, we use the multigrid V-cycle and the diagonal of the pressure mass matrix as FE sub-blocks in (28), fixing the choices

$$\widetilde{A} := \widetilde{A}_{\mathrm{mg}}, \quad \widetilde{S} := D_p. \tag{35}$$

5.2 *Stochastic Galerkin Matrices*

The SG preconditioners are also chosen according to complexity considerations and their spectral properties. In particular, we want to improve the condition of the SG matrices with respect to the discretization parameters k and M if possible. For the SG matrices based on a product of complete multivariate Hermite polynomials, there exist the following inclusion bounds [24, Corollary 4.5]:

$$-g_q \leq \frac{\boldsymbol{a}^{\mathrm{T}} G_q \boldsymbol{a}}{\boldsymbol{a}^{\mathrm{T}} \boldsymbol{a}} \leq g_q, \quad \forall \boldsymbol{a} \in \mathbb{R}^{Q_z} \backslash \{\mathbf{0}\}, \tag{36}$$

for $q = 1, \ldots, Q_\nu - 1$, where $g_q = \exp\left(M(k+1)/2 + \frac{1}{2} \sum_{m=1}^{M} k_m\right)$. As mentioned in Sect. 4 on page 69, the degrees k_m are the entries of the multi-index associated with the index q. According to the bounds (36), the eigenvalues of the SG matrices depend on the chaos degree k and the truncation index M of the KLE.

In the context of SGFE problems, the mean-based approximation [22] is often used to construct preconditioners for SG matrices. To define an SG preconditioner, the mean information of the SGFE problem is used. We work with an orthonormal chaos basis and the corresponding SG matrix of the mean problem is the identity matrix, see (22). According to our structural ansatz (28), we thus define

$$\widetilde{G}_A = \widetilde{G}_S = I. \tag{37}$$

We solely use this choice in the following although these preconditioners can not improve the condition of the SG matrices with respect to k or M. This is because of two reasons: the mean-based preconditioner is extremely cheap to apply and—to the best of our knowledge—there is no practical preconditioner which can eliminate the ill-conditioning with respect to k. There is still the potential ill-conditioning with respect to M. However, our numerical experiments in Sect. 8 suggest that the influence of M is not that severe.

6 Eigenvalue Analysis for the SGFE Matrices

In Sect. 5, we fixed the structures and building blocks of our preconditioners. As a next step, we summarize our assumptions and choices to define the specific preconditioners we eventually use. Starting point for the construction of the preconditioners are the structures (27) and the substructural Kronecker product ansatz (28).

Inserting $\widetilde{A} = \widetilde{A}_{\mathrm{mg}}$ and $\widetilde{S} = D_p$ as FE preconditioners, and $\widetilde{G}_A = \widetilde{G}_S = I$ as SG preconditioners, we define:

$$\mathcal{P}_{\mathrm{diag}} := \begin{bmatrix} \widetilde{\mathcal{A}}_{\mathrm{mg}} & 0 \\ 0 & \widetilde{\mathcal{S}}_p \end{bmatrix}, \quad \mathcal{P}_{\mathrm{tri}} := \begin{bmatrix} a\,\widetilde{\mathcal{A}}_{\mathrm{mg}} & 0 \\ \mathcal{B} & -\widetilde{\mathcal{S}}_p \end{bmatrix}, \tag{38}$$

where

$$\widetilde{\mathcal{A}}_{\mathrm{mg}} := I \otimes \widetilde{A}_{\mathrm{mg}}, \tag{39}$$

$$\widetilde{\mathcal{S}}_p := I \otimes D_p. \tag{40}$$

Both, the negative sign of the Schur complement approximation $-\widetilde{\mathcal{S}}_p$ as well as the scalar $a \in \mathbb{R}$ in the definition of the block triangular preconditioner $\mathcal{P}_{\mathrm{tri}}$ are manipulations which are not necessary in general. However, they are essential for one of the iterative solvers we are using. The specific effects of these additional manipulations are specified in Sect. 6.2.

In order to asses the influence of different problem parameters on the spectrum of the preconditioned SGFE systems, we want to derive inclusion bounds for the eigenvalues of $\mathcal{P}_{\mathrm{diag}}^{-1}\mathcal{C}$ and $\mathcal{P}_{\mathrm{tri}}^{-1}\mathcal{C}$ defined in (21) and (38). We can do this using existing results from saddle point theory. For the block diagonal preconditioned problem, we use [23, Theorem 3.2, Corollary 3.3 and Corollary 3.4]. These results imply that the eigenvalues of $\mathcal{P}_{\mathrm{diag}}^{-1}\mathcal{C}$ depend on the eigenvalues of—in our notation—$\widetilde{\mathcal{A}}_{\text{mg}}^{-1}\mathcal{A}$ and $\widetilde{\mathcal{S}}_p^{-1}\mathcal{B}\widetilde{\mathcal{A}}_{\text{mg}}^{-1}\mathcal{B}^{\text{T}}$. However, in order to apply these results, both preconditioned sub-matrices must be positive definite.

To derive estimates on the eigenvalues of the block triangular preconditioned matrix, we want to apply [32, Theorem 4.1]. This result bounds the eigenvalues of $\mathcal{P}_{\text{tri}}^{-1}\mathcal{C}$ by the eigenvalues of the preconditioned SGFE Laplacian and the preconditioned SGFE Schur complement. In our setting, it can be applied if

$$a\,\widetilde{\mathcal{A}}_{\text{mg}} < \mathcal{A} \leq \alpha_2 \widetilde{\mathcal{A}}_{\text{mg}}, \tag{41}$$

$$\hat{\gamma}\widetilde{\mathcal{S}}_p \leq \mathcal{B}\mathcal{A}^{-1}\mathcal{B}^{\text{T}} \leq \hat{\Gamma}\widetilde{\mathcal{S}}_p, \tag{42}$$

where a is the scaling introduced in (38) and α_2, $\hat{\gamma}$ and $\hat{\Gamma}$ are positive constants.

6.1 The Block Diagonal Preconditioned SGFE System

First of all, we consider the preconditioned SGFE Laplacian. To derive bounds on the eigenvalues of $\widetilde{\mathcal{A}}_{\text{mg}}^{-1}\mathcal{A}$, we proceed as in [19, Lemma 7.2], [24, Theorem 4.6].

Lemma 1 *Let the matrices $\mathcal{A}$ and $\widetilde{\mathcal{A}}_{mg}$ be defined according to* (22) *and* (39), *respectively. Then,*

$$\hat{\delta} \leq \frac{\boldsymbol{v}^T \mathcal{A}\, \boldsymbol{v}}{\boldsymbol{v}^T \widetilde{\mathcal{A}}_{mg}\, \boldsymbol{v}} \leq \hat{\Delta}, \quad \forall \boldsymbol{v} \in \mathbb{R}^{QN_u}\backslash\{\boldsymbol{0}\}, \tag{43}$$

with

$$\hat{\delta} := \underline{\delta} > 0, \tag{44}$$

$$\hat{\Delta} := (\overline{\nu}_0 + \nu_\sigma)\Delta, \tag{45}$$

where $\underline{\delta}$ is a positive constant not further specified and $\nu_\sigma := \sum_{q=1}^{Q_\nu - 1} g_q\, \overline{\nu}_q$.

Proof First of all, we bound the eigenvalues from above:

$$\lambda_{\max}\left(\widetilde{\mathcal{A}}_{\text{mg}}^{-1}\mathcal{A}\right) = \max_{\boldsymbol{v}\in\mathbb{R}^{QN_u}\backslash\{\boldsymbol{0}\}} \frac{\boldsymbol{v}^{\text{T}}\left(I \otimes A_0 + \sum_{q=1}^{Q_z-1} G_q \otimes A_q\right)\boldsymbol{v}}{\boldsymbol{v}^{\text{T}}\left(I \otimes \widetilde{A}_{\text{mg}}\right)\boldsymbol{v}} \tag{46}$$

$$\leq \lambda_{\max}\left(I \otimes \widetilde{A}_{\text{mg}}^{-1}A_0\right) + \sum_{q=1}^{Q_z-1} \lambda_{\max}\left(G_q \otimes \widetilde{A}_{\text{mg}}^{-1}A_q\right)$$

$$\overset{(31),(36)}{\leq} \overline{\nu}_0\,\Delta + \sum_{q=1}^{Q_\nu-1} g_q\,\overline{\nu}_q\,\Delta = (\overline{\nu}_0 + \nu_\sigma)\Delta. \tag{47}$$

An analog procedure for the lower bound on the eigenvalues yields

$$\lambda_{\min}\left(\widetilde{\mathcal{A}}_{\mathrm{mg}}^{-1}\mathcal{A}\right) \geq \underline{\nu}_0\,\underline{\delta} - \nu_\sigma\,\Delta. \tag{48}$$

However, due to the rough bounds g_q entering ν_σ, expression (48) is likely to be negative. This does not contradict the theory, but the results for the block diagonal preconditioned saddle point problem do not hold for a negative lower bound. The discrete well-posedness assumption ensures the existence of a positive lower bound $\lambda_{\min}(\mathcal{A}) \geq \underline{\alpha} > 0$ as it implies discrete coercivity. As $\widetilde{\mathcal{A}}_{\mathrm{mg}}^{-1}$ is positive definite as well, there is also a positive constant $\underline{\delta}$ fulfilling

$$\lambda_{\min}\left(\widetilde{\mathcal{A}}_{\mathrm{mg}}^{-1}\mathcal{A}\right) \geq \underline{\delta} > 0, \tag{49}$$

such that the result can be applied. However, we do not have any further information on $\underline{\delta}$, especially not on the parameter dependencies hidden in the bound. ■

Bounds on the eigenvalues of $\widetilde{\mathcal{S}}_p^{-1}\mathcal{B}\widetilde{\mathcal{A}}_{\mathrm{mg}}^{-1}\mathcal{B}^{\mathrm{T}}$ can be derived as in [19, Lemma 7.3]:

Lemma 2 ([19, Lemma 7.3]) *Let $\widetilde{\mathcal{A}}_{mg}$ and $\widetilde{\mathcal{S}}_p$ be defined as in* (39) *and* (40). *Then*

$$\underline{\delta}\,\theta\,\gamma^2 \leq \frac{\boldsymbol{q}^T\mathcal{B}\widetilde{\mathcal{A}}_{mg}^{-1}\mathcal{B}^T\boldsymbol{q}}{\boldsymbol{q}^T\widetilde{\mathcal{S}}_p\boldsymbol{q}} \leq \Delta\,\Theta, \quad \forall \boldsymbol{q} \in \mathbb{R}^{QN_p}\setminus\{\boldsymbol{0},\boldsymbol{1}\}, \tag{50}$$

Proof See proof of [19, Lemma 7.3]. ■

The constants in the bounds in (50) only depend on the degree and type of finite elements we use and on the shape of the considered domain. They do not depend on discretization or modeling parameters.

Combining Lemmas 1 and 2 with the results [23, Theorem 3.2, Corollary 3.3 and Corollary 3.4], we find that the eigenvalues of $\mathcal{P}_{\mathrm{diag}}^{-1}\mathcal{C}$ are bounded by a combination of (44), (45) and the bounds in (50). The bounds in (50) are parameter independent, so we look into (44) and (45). The bound (45) suggests that the eigenvalues change with the functions ν_q, $q = 0, \ldots, Q_z - 1$ and the parameters contained in the SG matrix bound g_q. Looking at the definition (36), the eigenvalues might thus be influenced by the chaos degree k and the KLE truncation index M. Lastly, we consider (44): As we do not have information on the parameter dependencies hidden in $\underline{\delta}$, any other problem parameter—such as the mesh width h—can potentially influence the eigenvalues.

6.2 The Block Triangular Preconditioned SGFE System

In order to bound the eigenvalues of $\mathcal{P}_{\mathrm{tri}}^{-1}\mathcal{C}$ using [32, Theorem 4.1], we need to fulfill (41) and (42). The right bound of (43) is an upper bound fulfilling (41) but the left bound of (43) is not necessarily a lower bound fulfilling (41). The lower bound in (41) implies in fact the stronger condition:

$$\frac{\boldsymbol{v}^{\mathrm{T}} \mathcal{A}\, \boldsymbol{v}}{\boldsymbol{v}^{\mathrm{T}} a \widetilde{\mathcal{A}}_{\mathrm{mg}}\, \boldsymbol{v}} > 1, \quad \forall \boldsymbol{v} \in \mathbb{R}^{QN_u} \backslash \{\mathbf{0}\}. \tag{51}$$

or equivalently

$$\lambda_{\min}\left((a\widetilde{\mathcal{A}}_{\mathrm{mg}})^{-1}\mathcal{A}\right) = \min_{\boldsymbol{v} \in \mathbb{R}^{QN_u} \backslash \{\mathbf{0}\}} \frac{\boldsymbol{v}^{\mathrm{T}} \mathcal{A} \boldsymbol{v}}{\boldsymbol{v}^{\mathrm{T}} a \widetilde{\mathcal{A}}_{\mathrm{mg}} \boldsymbol{v}} > 1. \tag{52}$$

Now, the scalar a comes into play. If it is chosen such that $a = \kappa\, \lambda_{\min}(\widetilde{\mathcal{A}}_{\mathrm{mg}}^{-1}\mathcal{A})$ with $0 < \kappa < 1$, then a scaled preconditioner $\kappa\, \lambda_{\min}(\widetilde{\mathcal{A}}_{\mathrm{mg}}^{-1}\mathcal{A})\, \widetilde{\mathcal{A}}_{\mathrm{mg}}$ fulfills (52) with

$$\lambda_{\min}\left((\kappa\, \lambda_{\min}(\widetilde{\mathcal{A}}_{\mathrm{mg}}^{-1}\mathcal{A})\, \widetilde{\mathcal{A}}_{\mathrm{mg}})^{-1}\mathcal{A}\right) = \kappa^{-1} > 1. \tag{53}$$

However, as we have no quantitative access to the analytical lower bound (44), we need to estimate the minimum eigenvalue of $\widetilde{\mathcal{A}}_{\mathrm{mg}}^{-1}\mathcal{A}$ numerically. A scaled preconditioner with a numerically computed positive $a = a^* < \lambda_{\min}(\widetilde{\mathcal{A}}_{\mathrm{mg}}^{-1}\mathcal{A})$ then yields a modified version of (43), namely

$$1 < \frac{\hat{\delta}}{a^*} \leq \frac{\boldsymbol{v}^{\mathrm{T}} \mathcal{A}\, \boldsymbol{v}}{\boldsymbol{v}^{\mathrm{T}} a^*\, \widetilde{\mathcal{A}}_{\mathrm{mg}}\, \boldsymbol{v}} \leq \frac{\hat{\Delta}}{a^*}, \quad \forall \boldsymbol{v} \in \mathbb{R}^{QN_u} \backslash \{\mathbf{0}\}, \tag{54}$$

which fulfills (41).

As we have assumed well-posedness of the discrete SGFE problem, a discrete inf-sup condition is fulfilled [17, Theorem 3.18]. Using the discrete representations of the norms and inner products in the FE and SG basis, we can rearrange the discrete inf-sup condition to derive a relation exactly as in [17, Lemma 3.48 and section 3.6.6]. In our SGFE setting, the spectral equivalence has the form

$$\beta^2 \leq \frac{\boldsymbol{q}^{\mathrm{T}}(I \otimes BA^{-1}B^{\mathrm{T}})\boldsymbol{q}}{\boldsymbol{q}^{\mathrm{T}}(I \otimes M_p)\boldsymbol{q}} \leq 1, \quad \forall \boldsymbol{q} \in \mathbb{R}^{QN_p} \backslash \{\mathbf{0}, \mathbf{1}\}. \tag{55}$$

Here, β is the inf-sup constant of the mixed SGFE problem. We use (55) to derive bounds for the preconditioned SGFE Schur complement in the following lemma.

Lemma 3 *Let the Schur complement be defined by* $\mathcal{S} = -\mathcal{B}\mathcal{A}^{-1}\mathcal{B}^T$ *with building blocks* $\mathcal{A}$ *and* $\mathcal{B}$ *in* (22) *and* (23), *and* $\widetilde{\mathcal{S}}_p = I \otimes D_p$ *according to* (40). *Then,*

$$\hat{\gamma} \leq \frac{\boldsymbol{q}^T \mathcal{B}\mathcal{A}^{-1}\mathcal{B}^T \boldsymbol{q}}{\boldsymbol{q}^T \left(I \otimes D_p\right) \boldsymbol{q}} \leq \hat{\Gamma}, \quad \boldsymbol{q} \in \mathbb{R}^{QN_p} \backslash \{\mathbf{0}, \mathbf{1}\}, \tag{56}$$

with

$$\hat{\gamma} := (\underline{\nu}_0 + \nu_\sigma)^{-1} \theta\, \beta^2, \tag{57}$$

$$\hat{\Gamma} := \underline{\delta}^{-1}\, \Delta\, \Theta. \tag{58}$$

Proof We want to modify (55) in order to derive (56). The denominators are matched directly via (33) as the additional identity matrices do not change the bounds. The connection between the nominators is not that obvious. We start by considering

$$(I \otimes A)^{-1} \mathcal{A} = I \otimes A^{-1} A_0 + \sum_{q=1}^{Q_\nu - 1} G_q \otimes A^{-1} A_q. \tag{59}$$

Extracting the weighting factors from the FE Laplacians yields

$$-\,\overline{\nu}_q \leq \frac{\boldsymbol{v}^{\mathrm{T}} A_q \boldsymbol{v}}{\boldsymbol{v}^{\mathrm{T}} A\, \boldsymbol{v}} \leq \overline{\nu}_q, \tag{60}$$

for all $q = 1, \ldots Q_\nu - 1$ and for all $\boldsymbol{v} \in \mathbb{R}^{N_u} \backslash \{\mathbf{0}\}$. By combining (60) with the representation (59) and (36), we get

$$\frac{\boldsymbol{v}^{\mathrm{T}} (I \otimes A)^{-1} \boldsymbol{v}}{\boldsymbol{v}^{\mathrm{T}} \mathcal{A}^{-1} \boldsymbol{v}} \leq \overline{\nu}_0 + \sum_{q=1}^{Q_\nu - 1} g_q\, \overline{\nu}_q = \overline{\nu}_0 + \nu_\sigma, \tag{61}$$

for all $\boldsymbol{v} \in \mathbb{R}^{QN_u} \backslash \{\mathbf{0}\}$. For the lower bound, we start with (43):

$$\frac{\boldsymbol{v}^{\mathrm{T}} \mathcal{A}\, \boldsymbol{v}}{\boldsymbol{v}^{\mathrm{T}} \widetilde{\mathcal{A}}_{\mathrm{mg}}\, \boldsymbol{v}} \geq \underline{\delta}, \quad \forall \boldsymbol{v} \in \mathbb{R}^{QN_u} \backslash \{\mathbf{0}\}. \tag{62}$$

As $\widetilde{\mathcal{A}}_{\mathrm{mg}} = I \otimes \widetilde{A}_{\mathrm{mg}}$, we can use the inverse of (30) with (62) and derive

$$\frac{\boldsymbol{v}^{\mathrm{T}} (I \otimes A)^{-1} \boldsymbol{v}}{\boldsymbol{v}^{\mathrm{T}} \mathcal{A}^{-1} \boldsymbol{v}} \geq \underline{\delta}\, \Delta^{-1}, \quad \forall \boldsymbol{v} \in \mathbb{R}^{QN_u} \backslash \{\mathbf{0}\}. \tag{63}$$

Now, we use the relation $\boldsymbol{v} = \mathcal{B}^{\mathrm{T}} \boldsymbol{q}$ in (61) and (62), to establish:

$$(\underline{\nu}_0 + \nu_\sigma)^{-1} \leq \frac{\boldsymbol{q}^{\mathrm{T}} \mathcal{B} \mathcal{A}^{-1} \mathcal{B}^{\mathrm{T}} \boldsymbol{q}}{\boldsymbol{q}^{\mathrm{T}} \mathcal{B} (I \otimes A)^{-1} \mathcal{B}^{\mathrm{T}} \boldsymbol{q}} \leq \underline{\delta}^{-1}\, \Delta, \tag{64}$$

for all $\boldsymbol{q} \in \mathbb{R}^{QN_p} \backslash \{\mathbf{0}, \mathbf{1}\}$. Multiplying the positive expressions (55) and (64) and using (33) then yields the assertion. ■

Using (54) and Lemma 3 with the result [32, Theorem 4.1] implies that the eigenvalues of $\mathcal{P}_{\text{tri}}^{-1}\mathcal{C}$ with $a = a^* < \lambda_{\min}(\widetilde{\mathcal{A}}_{\text{mg}}^{-1}\mathcal{A})$ can be bounded by a combination of the bounds in (54), (57) and (58). The upper bound in (54) and (57) suggest that the eigenvalues are influenced by the scaling a^*, the chaos degree k and the KLE truncation index M hidden in the bounds g_q, and the functions ν_q. However, we do not know which parameters are hidden in (44) and (58). Therefore, we can not exclude the possibility that the eigenvalues change with other problem parameters such as the mesh width h.

We now introduce the matrix

$$\mathcal{H} := \begin{bmatrix} \mathcal{A} - a\widetilde{\mathcal{A}}_{\text{mg}} & 0 \\ 0 & \widetilde{\mathcal{S}}_p \end{bmatrix} \tag{65}$$

with $a = \kappa\, \lambda_{\min}(\widetilde{\mathcal{A}}_{\text{mg}}^{-1}\mathcal{A})$, $0 < \kappa < 1$, $\widetilde{\mathcal{A}}_{\text{mg}}$ and $\widetilde{\mathcal{S}}_p$ according to (39) and (40). We need this matrix in the following lemma, which can be proven because of the modifications to the block triangular preconditioner $\mathcal{P}_{\text{tri}}$ in (38).

Lemma 4 *Let a in* (38) *be set to $a = \kappa\, \lambda_{min}(\widetilde{\mathcal{A}}_{mg}^{-1}\mathcal{A})$, with $0 < \kappa < 1$. Then, the matrix $\mathcal{H}$ in* (65) *defines an inner product and the triangular preconditioned system matrix $\mathcal{P}_{tri}^{-1}\mathcal{C}$ with $\mathcal{C}$ and $\mathcal{P}_{tri}$ according to* (21) *and* (38)*, is $\mathcal{H}$-symmetric and $\mathcal{H}$-positive definite, i.e.*

$$\mathcal{H}\mathcal{P}_{tri}^{-1}\mathcal{C} = (\mathcal{P}_{tri}^{-1}\mathcal{C})^T\mathcal{H}, \tag{66}$$

$$z^T\mathcal{H}\mathcal{P}_{tri}^{-1}\mathcal{C}z > 0, \qquad \forall z \in \mathbb{R}^{QN}\backslash\{\mathbf{0}\}. \tag{67}$$

Proof See proof of [19, Lemma 7.5]. ■

Bramble and Pasciak discovered the extraordinary effect of the matrix $\mathcal{H}$ in their 1988 paper [7]. They considered a triangular preconditioned problem structurally equivalent to $\mathcal{P}_{\text{tri}}^{-1}\mathcal{C}$. Due to the conditions verified in Lemma 4, a conjugate gradient method exists in the $\mathcal{H}\mathcal{P}_{\text{tri}}^{-1}\mathcal{C}$-inner product [12]. The Bramble-Pasciak conjugate gradient (BPCG) method introduced in [7] can thus also be applied to the SGFE problem considered in this work.

7 Iterative Solvers

In the following, we discuss two different iterative solvers for the SGFE Stokes problem. Similar investigations were conducted in [21] for Stokes flow with deterministic data and in [19] for Stokes flow with uniform random data.

As the system matrix $\mathcal{C}$ of the SGFE Stokes problem (20) is symmetric but indefinite (see (26)), the MINRES method is the first iterative solver one usually considers. It is attractive from a complexity point of view as the Krylov basis is built with short recurrence. However, MINRES relies on the symmetry of the problem and can thus only be combined with a symmetric preconditioner. Applying e.g. the nonsymmetric $\mathcal{P}_{\text{tri}}$ in (38) may prevent MINRES from converging. Concerning the preliminary considerations in the Sects. 5 and 6, we will thus use MINRES solely in combination with $\mathcal{P}_{\text{diag}}$ in (38). The MINRES algorithm can be formulated efficiently such that the only matrix-vector operations necessary per iteration are the application of $\mathcal{C}$ and $\mathcal{P}_{\text{diag}}^{-1}$, see [8, Algorithm 4.1]. Bounds on the extreme eigenvalues of $\mathcal{P}_{\text{diag}}^{-1}\mathcal{C}$ are often used to asses MINRES convergence behavior a priori. This is done because of the standard convergence result [8, Theorem 4.14] which bounds MINRES iteration numbers by a function of those eigenvalues. Following this reasoning, we will use the results from Sect. 6.1 to interpret the numerical MINRES results in Sect. 8.

The SGFE problem is no longer symmetric when $\mathcal{P}_{\text{tri}}^{-1}$ in (38) is applied to $\mathcal{C}$. Faber and Manteuffel [12] proofed that there does not exist a short recurrence for generating an orthogonal Krylov subspace basis for every nonsymmetric matrix. However, they showed that there are special cases for which it is possible. The combination of $\mathcal{P}_{\text{tri}}^{-1}\mathcal{C}$ and $\mathcal{H}$ is such a case, as condition (66) holds. Systems of equations associated with the nonsymmetric matrix $\mathcal{P}_{\text{tri}}^{-1}\mathcal{C}$ can thus be solved with a CG method. Besides this big advantage, there are also drawbacks: firstly, for the method to be defined properly, we must scale $\widetilde{\mathcal{A}}_{\text{mg}}$ such that (52) holds. This can be achieved by choosing the scaling to be $a = \kappa\, \lambda_{\min}(\widetilde{\mathcal{A}}_{\text{mg}}^{-1}\mathcal{A})$, $0 < \kappa < 1$, as discussed in Sect. 6.2. However, solving the associated eigenproblem numerically leads to additional computational costs. Secondly, the naive BPCG algorithm is associated with the $\mathcal{H}\mathcal{P}_{\text{tri}}^{-1}\mathcal{C}$-inner product [1, section 4]. Evaluating quantities in this inner product would lead to additional matrix-vector operations. Due to certain properties of CG methods [1], these additional costs can be avoided by reformulating the algorithm. Thereby, a BPCG algorithm can be found which needs only one extra operation compared to preconditioned MINRES [21, section 3.1]: a matrix-vector multiplication with $\mathcal{B}$. This additional operation originates from the definition of $\mathcal{P}_{\text{tri}}$ in (38) and is cheap compared to a multiplication with $\mathcal{A}$, because $\mathcal{B}$ is block diagonal with sparse blocks. For this reason, the BPCG method is particularly interesting in our setting where $\mathcal{A}$ is block dense [10, Lemma 28]. There is also a convergence result for CG which bounds its iteration numbers by a function of the extreme eigenvalues of $\mathcal{P}_{\text{tri}}^{-1}\mathcal{C}$ [15, Theorem 9.4.12], so we follow the same arguments as above: We use the results from Sect. 6.2 to interpret the numerical CG results in Sect. 8.

8 Numerical Experiments

In the following, we compare the solvers discussed in Sect. 7 numerically. We use the regularized driven cavity test case and investigate the performance of the two methods as well as their convergence behavior.

We associate the random field $\mu(x, \omega)$ with the separable exponential covariance function $C_\mu : D \times D \to \mathbb{R}$:

$$C_\mu(x, y) = \sigma_\mu^2 e^{-|x_1-y_1|/b_1-|x_2-y_2|/b_2}, \tag{68}$$

with correlation lengths b_1 and b_2 in the x_1 and x_2 direction, respectively. Eigenpairs of the two-dimensional integral operator associated with (68) are constructed by combining the eigenpairs of two one-dimensional operators, which can be calculated analytically [13, section 5.3], [18, section 7.1].

We use a regularized version of the lid-driven cavity [8, section 3.1] as our test case. The spatial domain is the unit square $D = [-0.5, 0.5] \times [-0.5, 0.5]$ and we impose a parabolic flow profile $u(x) = (1 - 16x_1^4, 0)^T$ at the top lid. No-slip conditions are enforced everywhere else on the boundary. For the numerical simulations, we use the following default parameter set:

$$h = 0.01, \quad k = 1, \quad M = 10, \quad \nu_0 = 1, \quad \sigma_\nu = 0.2, \quad b_1 = b_2 = 1. \tag{69}$$

If not specified otherwise, the simulation parameters are the ones in (69). The mean and variance of the corresponding velocity streamline field can be found in Fig. 1. All numerical simulations are carried out in our own finite element implementation in MATLAB [28], except the setup of the multigrid preconditioner. For this particular issue, we resort to the algebraic multigrid implementation in the IFISS package [9] with two point Gauss-Seidel pre-and post-smoothing sweeps. In order to compare $\mathcal{P}_{\text{diag}}$-preconditioned MINRES ($\mathcal{P}_{\text{diag}}$-MINRES) and $\mathcal{P}_{\text{tri}}$-preconditioned BPCG ($\mathcal{P}_{\text{tri}}$-BPCG), we look at the iteration counts necessary to reduce the Euclidean norm of the relative residual below 10^{-6}. As the initial guess is always the zero vector, we thus consider numbers n such that

$$\|\boldsymbol{r}^{(n)}\| = \|\boldsymbol{b} - \mathcal{C}\boldsymbol{z}^{(n)}\| \leq 10^{-6}\|\boldsymbol{b}\|. \tag{70}$$

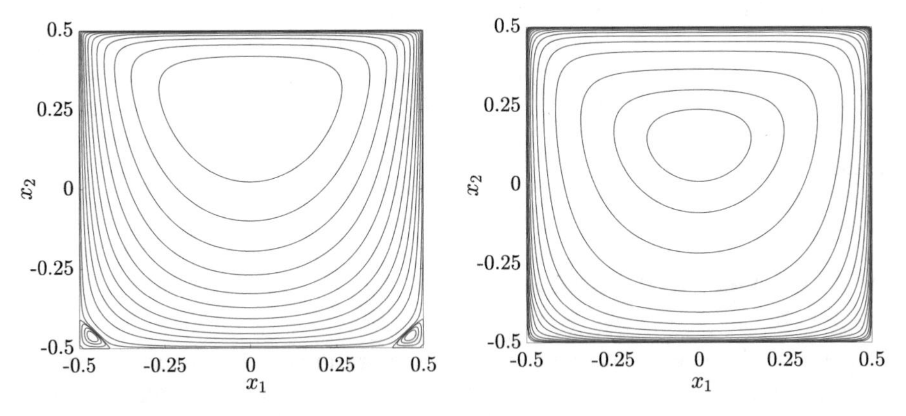

Fig. 1 Contour lines of the mean (left) and variance (right) of the stream function of the regularized driven-cavity test case computed with the parameters in (69)

Table 1 $\mathcal{P}_{\text{tri}}$-BPCG iteration counts for different values of the relative scaling a/a^*

a/a^*	0.1	0.4	0.6	0.8	0.9	1.0	1.1	1.2	1.4	1.6	2.0	3.0	5.0
n	46	41	39	35	34	32	35	35	35	36	38	41	42

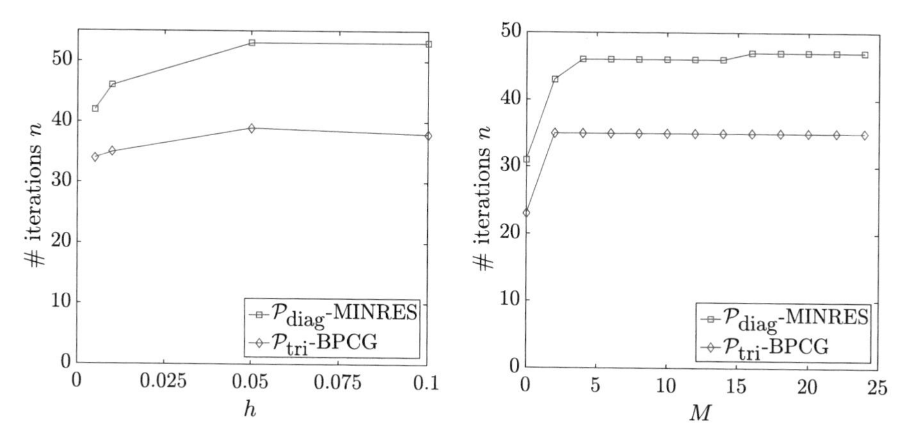

Fig. 2 Iteration counts for different values of the mesh size h (left) and the KLE truncation index M (right) for $\mathcal{P}_{\text{diag}}$-MINRES (red) and $\mathcal{P}_{\text{tri}}$-BPCG (blue)

Before we commence with the actual comparison, we want to assess the influence of the scaling factor a on the $\mathcal{P}_{\text{tri}}$-BPCG convergence. Based on the reference value $a^* \approx \lambda_{\min}(\widetilde{\mathcal{A}}_{\text{mg}}^{-1}\mathcal{A})$ that we compute with MATLAB's (version 8.6.0) numerical eigensolver `eigs`, we solve the driven cavity problem for different values of the relative scaling a/a^*. Corresponding iteration counts for $\mathcal{P}_{\text{tri}}$-BPCG are displayed in Table 1. The minimum iteration count of 32 is attained when the scaling a is chosen to be the reference value a^*. Consequently, the ideal scaling of the preconditioner is close to the border of the $\mathcal{H}$-positive definiteness condition (52). However, we notice that moderate variations around the optimal scaling do not lead to a significant increase in iteration counts. Further, we can not guarantee convergence of the algorithm when $a/a^* > 1$ as the $\mathcal{H}$-positive definiteness requirement (52) is no longer strictly fulfilled. Still, we could not observe divergent behavior in our experiments.

To lower the costs associated with computing a^*, we solve the associated eigenproblems on the coarsest mesh with $h = 0.1$. This is somewhat heuristic as we could not show h-independence of the bounds in Lemma 1. Nevertheless, we assume that the mesh size does not influence the scaling significantly due to the chosen spectrally equivalent FE preconditioners, see Sect. 5.1.

We now look at the iteration counts of the two solvers when different parameters are varied, starting with the mesh size h and the truncation index M of the KLE. The associated numerical results are displayed in Fig. 2. First of all, we observe that $\mathcal{P}_{\text{tri}}$-BPCG converges after fewer iterations than $\mathcal{P}_{\text{diag}}$-MINRES for all considered values of h and M. Further, the iteration counts of both solvers do not increase under mesh refinement but rather decrease slightly, as can be seen in the left plot. In the right plot, we notice that the iteration counts increase up to $M \approx 5$ for both methods and then basically stay constant independent of M. The results in Fig. 2 suggest that the iteration numbers are asymptotically independent of the mesh size and the KLE truncation index. This is according to expectations for h, as the multigrid V-cycle and the diagonal of the pressure mass matrix are spectrally equivalent to the weighted FE Laplacians and the Schur complement, see (30)–(34). Asymptotic independence of M is somewhat surprising, as this parameter appears in (45), hidden in ν_σ. This dependence originates from the bounds on the SG matrices in (36).

Figure 3 visualizes the convergence behavior of the two considered solvers when either the total degree k of the chaos basis or the standard deviation σ_μ of the original Gaussian process $\mu(x, \omega)$ in (2) is varied. We again observe that $\mathcal{P}_{\text{diag}}$-MINRES consistently needs more iterations to converge than $\mathcal{P}_{\text{tri}}$-BPCG. We can further see a steady increase of iteration counts with both k and σ_μ for both solvers. That was to be expected as these parameters also occur in the bounds (45) and (57). When they increase, the fluctuation parts—i.e. the terms in the sum in (22)—become more important, see (17) and (36).

In order to alleviate the influence on k and σ_μ, one needs to use more advanced approaches for the SG preconditioners such as the Kronecker product preconditioner, see [24, 27]. However, as there is—to the best of our knowledge—no practical preconditioner that can eliminate just one of these dependencies and using a more elaborate preconditioner also results in increased computational costs, we do not investigate this issue further here.

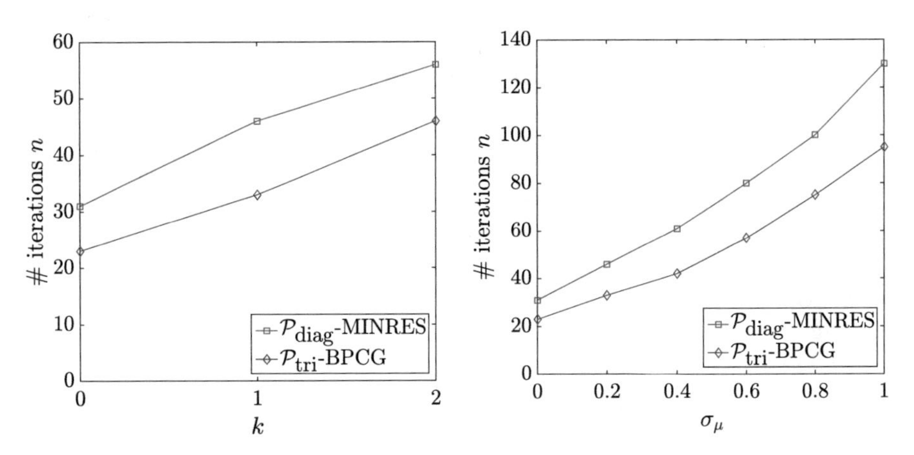

Fig. 3 Iteration counts for different values of the chaos degree k (left) and the standard deviation σ_ν (right) of the Gaussian field $\mu(x, \omega)$ for $\mathcal{P}_{\text{diag}}$-MINRES (red) and $\mathcal{P}_2$-BPCG (blue)

9 Conclusion

The construction of our BPCG solver relies on the appropriate choice of the scaling a such that the matrix $\mathcal{H}$ is positive definite. Choosing a close to the minimum eigenvalue of the preconditioned SGFE Laplacian is optimal, as confirmed by the numerical experiments. However, solving the associated eigenproblem numerically is often prohibitive. To reduce the costs of computing a, one can solve the eigenproblem on a coarser mesh. This approach worked well in the numerical examples we considered.

We compared the iteration counts of two iterative solvers with structurally different preconditioners: block diagonal preconditioned MINRES and block triangular preconditioned BPCG. The iteration numbers of the latter were consistently lower in our experiments. However, as we used the same FE and SG building blocks, the performance was qualitatively the same: The application of the multigrid preconditioner and the diagonal of the pressure mass matrix resulted in iteration numbers basically independent of the mesh width h. The input dimension M is a measure for the accuracy of the input representation. However, as soon as a certain threshold is reached, the iteration numbers stayed constant independent of M. This suggests that the eigenvalues of the SG matrices are asymptotically independent of M, an assertion which is not according to the derived eigenvalue bounds. Both the degree of the polynomial chaos k as well as the standard deviation σ_μ critically influence the condition of our preconditioned problems. This is already visible in the available eigenvalue bounds. The mean-based SG preconditioner can not alleviate these influences. Therefore, when we increased one of these parameters, iteration numbers increased as well.

Summarizing the investigations of the BPCG method with block triangular preconditioner and the MINRES method with block diagonal preconditioner, we can state the following: The eigenvalue analysis is largely inconclusive mainly due to the coarse inclusion bounds for the SG matrices. However, our numerical tests suggest that the methods perform similarly to each other and essential behave as expected. If the scaling parameter can be obtained cheaply, the application of the block triangular preconditioner can result in a significant reduction of iteration counts compared to the application of the block diagonal preconditioner. This leads to a major reduction of the overall computational costs, because one step of block diagonal preconditioned MINRES is—especially in the SGFE case—only marginally cheaper than one step of block triangular preconditioned BPCG.

Acknowledgements This work is supported by the 'Excellence Initiative' of the German federal and state governments and the Graduate School of Computational Engineering at Technische Universität Darmstadt.

References

1. Ashby, S.F., Manteuffel, T.A., Saylor, P.E.: A taxonomy for conjugate gradient methods. SIAM J. Numer. Anal. **27**(6), 1542–1568 (1990). https://doi.org/10.1137/0727091
2. Babuška, I., Tempone, R., Zouraris G.E.: Galerkin finite element approximations of stochastic elliptic partial differential equations. SIAM J. Numer. Anal. **42**(2), 800–825 (2004). https://doi.org/https://doi.org/10.1137/S0036142902418680
3. Babuška, I., Nobile, F., Tempone, R.: A stochastic collocation method for elliptic partial differential equations with random input data. SIAM J. Numer. Anal. **45**(3), 1005–1034 (2007). https://doi.org/10.1137/050645142
4. Bachmayr, M., Cohen, A., DeVore, R., Migliorati, G.: Sparse polynomial approximation of parametric elliptic PDEs. Part II: lognormal coefficients. ESAIM: M2AN **51**(1), 341–363 (2017). https://doi.org/10.1051/m2an/2016051
5. Benzi, M., Golub, G.H., Liesen, J.: Numerical solution of saddle point problems. Acta Numer. **14**, 1–137 (2005). https://doi.org/10.1017/S0962492904000212
6. Bespalov, A., Powell, C.E., Silvester, D.: A priori error analysis of stochastic Galerkin mixed approximations of elliptic PDEs with random data. SIAM J. Numer. Anal. **50**(4), 2039–2063 (2012). https://doi.org/10.1137/110854898
7. Bramble, J.H., Pasciak, J.E.: A preconditioning technique for indefinite systems resulting from mixed approximations of elliptic problems. Math. Comput. **50**(181), 1–17 (1988). https://doi.org/10.1090/S0025-5718-1988-0917816-8
8. Elman, H.C., Silvester, D.J., Wathen, A.J.: Finite Elements and Fast Iterative Solvers: With Applications in Incompressible Fluid Dynamics, 2nd edn. Oxford University Press, Oxford (2014)
9. Elman, H.C., Ramage, A., Silvester, D.J.: IFISS: A computational laboratory for investigating incompressible flow problems. SIAM Rev. **56**(2), 261–273 (2014). https://doi.org/10.1137/120891393
10. Ernst, O.G., Ullmann, E.: Stochastic Galerkin matrices. SIAM J. Matrix Anal. Appl. **31**(4), 1848–1872 (2010). https://doi.org/10.1137/080742282
11. Ernst, O.G., Powell, C.E., Silvester, D.J., Ullmann, E.: Efficient solvers for a linear stochastic Galerkin mixed formulation of diffusion problems with random data. SIAM J. Sci. Comput. **31**(2), 1424–1447 (2009). https://doi.org/10.1137/070705817
12. Faber, V., Manteuffel, T.: Necessary and sufficient conditions for the existence of a conjugate gradient method. SIAM J. Numer. Anal. **21**(2), 352–362 (1984). https://doi.org/10.1137/0721026
13. Ghanem, R.G., Spanos, P.D.: Stochastic Finite Elements—A Spectral Approach. Springer, New York (1991)
14. Gunzburger, M.D., Webster, C.G., Zhang, G.: Stochastic finite element methods for partial differential equations with random input data. Acta Numer. **23**, 521–650 (2014). https://doi.org/10.1017/S0962492914000075
15. Hackbusch, W.: Iterative Solution of Large Sparse Systems of Equations, 1st edn. Springer, New York (1994)
16. Hoang, V.H., Schwab, C.: N-term Wiener chaos approximation rates for elliptic PDEs with lognormal Gaussian random inputs. Math. Models Methods Appl. Sci. **24**(4), 797–826 (2014). https://doi.org/10.1142/S0218202513500681
17. John, V.: Finite Element Methods for Incompressible Flow Problems. Springer International Publishing, Cham (2016)
18. Lord, G.J., Powell, C.E., Shardlow, R.: An Introduction to Computational Stochastic PDEs. Cambridge University Press, New York (2014)
19. Müller, C., Ullmann, S., Lang, J.: A Bramble-Pasicak conjugate gradient method for discrete Stokes equations with random viscosity. Preprint. arXiv:1801.01838 (2018). https://arxiv.org/abs/1801.01838

20. Oksendal, B.: Stochastic Differential Equations: An Introduction with Applications, 5th edn. Springer, Berlin (1998)
21. Peters, J., Reichelt, V., Reusken, A.: Fast iterative solvers for discrete Stokes equations, SIAM J. Sci. Comput. **27**(2), 646–666 (2005). https://doi.org/10.1137/040606028
22. Powell, C.E., Elman, H.C.: Block-diagonal preconditioning for spectral stochastic finite-element systems. IMA J. Numer. Anal. **29**(2), 350–375 (2009). https://doi.org/10.1093/imanum/drn014
23. Powell, C.E., Silvester, D.: Optimal preconditioning for Raviart–Thomas mixed formulation of second-order elliptic problems. SIAM J. Matrix Anal. Appl. **25**(3), 718–738 (2003). https://doi.org/10.1137/S0895479802404428
24. Powell, C.E., Ullmann, E.: Preconditioning stochastic Galerkin saddle point systems. SIAM J. Matrix Anal. Appl. **31**(5), 2813–2840 (2010). https://doi.org/10.1137/090777797
25. Schwab, C., Gittelson, C. J.: Sparse tensor discretizations of high-dimensional parametric and stochastic PDEs. Acta Numer. **20**, 291–467 (2011). https://doi.org/10.1017/S0962492911000055
26. Ullmann, E.: Solution strategies for stochastic finite element discretizations. PhD thesis, Bergakademie Freiberg University of Technology (2008)
27. Ullmann, E.: A Kronecker product preconditioner for stochastic Galerkin finite element discretizations. SIAM J. Sci. Comput. **32**(2), 923–946 (2010). https://doi.org/10.1137/080742853
28. Ullmann S.: Triangular Taylor Hood finite elements, version 1.4. Retrieved: 06 October 2017. www.mathworks.com/matlabcentral/fileexchange/49169
29. Ullmann, E., Elman, H.C., Ernst, O.G.: Efficient iterative solvers for stochastic Galerkin discretizations of log-transformed random diffusion problems. SIAM J. Sci. Comput. **34**(2), A659–A682 (2012). https://doi.org/10.1137/110836675
30. Wathen, A.J.: On relaxation of Jacobi iteration for consistent and generalized mass matrices. Commun. Appl. Numer. Methods **7**(2), 93–102 (1991). https://doi.org/10.1002/cnm.1630070203
31. Xiu, D., Karniadakis, G.E.: The Wiener–Askey polynomial chaos for stochastic differential equations. SIAM J. Sci. Comput. **24**(2), 619–644 (2002). https://doi.org/10.1137/S1064827501387826
32. Zulehner, W.: Analysis of iterative methods for saddle point problems: a unified approach. Math. Comput. **71**(238), 479–505 (2001). http://www.jstor.org/stable/2698830

Linear Solvers for the Finite Pointset Method

Fabian Nick, Bram Metsch, and Hans-Joachim Plum

Abstract Many simulations in Computational Engineering suffer from slow convergence rates of their linear solvers. This is also true for the Finite Pointset Method (FPM), which is a Meshfree Method used in Computational Fluid Dynamics. FPM uses Generalized Finite Difference Methods (GFDM) in order to discretize the arising differential operators. Like other Meshfree Methods, it does not involve a fixed mesh; FPM uses a point cloud instead. We look at the properties of linear systems arising from GFDM on point clouds and their implications on different types of linear solvers, specifically focusing on the differences between one-level solvers and Multigrid Methods, including Algebraic Multigrid (AMG). With the knowledge about the properties of the systems, we develop a new Multigrid Method based on point cloud coarsening. Numerical experiments show that our Multicloud method has the same advantages as other Multigrid Methods; in particular its convergence rate does not deteriorate when refining the point cloud. In future research, we will examine its applicability to a broader range of problems and investigate its advantages in terms of computational performance.

Keywords Point cloud · Generalized finite difference methods · Multigrid

1 Introduction

One of the drawbacks of Finite Element Methods, Finite Volume Methods and other classical methods based on meshes is the need to generate a mesh. This becomes even more critical if the domain changes over the course of the simulation, for example when dealing with moving geometries or free surfaces. Meshfree Methods drop the idea of using a fixed mesh. There are two main classes of Meshfree Methods: Strong Form Meshfree Methods and Weak Form Meshfree Methods, see

---

F. Nick (✉) · B. Metsch · H.-J. Plum
Fraunhofer SCAI, Sankt Augustin, Germany
e-mail: fabian.nick@scai.fraunhofer.de; http://www.scai.fraunhofer.de

M. Schäfer et al. (eds.), *Recent Advances in Computational Engineering*, Lecture Notes in Computational Science and Engineering 124,
https://doi.org/10.1007/978-3-319-93891-2_6

[1] for an extensive overview. Our work focuses on a special Strong Form Meshfree Method called Finite Pointset Method (FPM) and specifically on the aspect of how to apply the idea of Multigrid Methods [18] in order to solve the linear systems arising from this method efficiently, cf. [5]. The aspect of solving the linear systems is crucial because with the increasing demand for accuracy of a simulation and therefore increasingly fine discretizations, classical iterative linear solvers suffer from their characteristic growth of the number of iterations needed to solve the linear systems, see Fig. 5. Multigrid Methods like Algebraic Multigrid (AMG) on the other hand show an almost constant number of iterations independently of the problem size. FPM uses Generalized Finite Difference Methods (GFDM) in order to discretize various differential operators, therefore this paper will focus on the linear systems produced by GFDM discretizations of the Laplace operator, which plays an important role in FPM.

2 Generalized Finite Difference Methods on Point Clouds

When we need to simulate processes with characteristics like free surfaces or moving geometries, we want to drop the idea of having a mesh. Instead, we aim to find means to define Finite Differences on unstructured sets of points $\mathcal{P} = \left\{\mathbf{x}_i \in \Omega \subset R^d\right\}, i = 1, \ldots, n$, which we will call *point clouds*, in order to solve an elliptical boundary value problem

$$\begin{cases} \mathcal{L}u = f & \text{on } \Omega \\ \mathcal{B}u = g & \text{on } \partial\Omega \end{cases} . \tag{1}$$

In this paper, we will only be dealing with the case $d = 2$.

2.1 Least Squares Methods

Given a point cloud $\mathcal{P}$ discretizing the domain Ω, we are interested in stencils $\mathfrak{s}_i^* = \left(c_{ij}^*\right)$ such that

$$u^* \approx \sum_{j \in N_i} c_{ij}^* u(\mathbf{x}_j) , \tag{2}$$

where N_i is a set of indices defining a neighborhood of $\mathbf{x}_i$, including $\mathbf{x}_i$. The superscript $*$ indicates any differential operator that is to be approximated. This can be the Laplace operator Δ for example. In order to find an approximation (2), Least Squares methods can be employed. Tiwari and Kuhnert [16, 17] follow a Moving

Least Squares idea [3]: In their method, the neighborhood N_i in (2) is chosen to consist of all points $\mathbf{x}_j$ within a ball of radius βh around $\mathbf{x}_i$. Note how their approach is different from similar methods like Smoothed Particle Hydrodynamics (SPH) [13] in that the points in the point cloud are only used to set up differential operators and do not carry any mass or other physical properties.

Definition 1 The parameter h is called *smoothing length* and is closely related to the *discretization size* h in standard Finite Difference Methods (FDM) on uniform grids. It is the radius of the support of the weighting function w we will introduce later. Like in Finite Difference or Finite Element Methods, the smoothing length can also be varied locally. We then write $h(\mathbf{x}_i)$, indicating the radius of the support of the weighting function at the point $\mathbf{x}_i$. The parameter β is used to control the number of neighboring points independently of the size of the support. Although all values $\beta \in [0, 1]$ are admissible, a standard choice is $\beta \in [0.75, 1]$ [15].

Taylor expansion yields that in order to obtain an exact approximation (2) for monomials u up to a certain order (usually order two), the coefficients c_{ij}^* have to satisfy M conditions that can be written as a linear system of the form

$$K_i^T \mathfrak{s}_i^* = \mathbf{b}^* \,, \tag{3}$$

where

$$K_i^T = \left((\mathbf{k}_i^m)^T \right)_{1 \leq m \leq M} \,, \; \mathbf{k}_i^m \in \mathbb{R}^{|N_i|} \,, \tag{4}$$

$$\mathfrak{s}_i^* = \left(c_{ij}^* \right)_{1 \leq j \leq |N_i|} \,, \tag{5}$$

$$\mathbf{b}^* = \left(\mathbf{b}_m^* \right)_{1 \leq m \leq M} \,. \tag{6}$$

Geometric conditions on the positions of the points $\mathbf{x}_j \in N_i$ would be necessary in order to guarantee a solution to the linear system (3), cf. [4, 7]. As checking and satisfying those conditions would become rather involved, especially if the point cloud $\mathcal{P}$ is changing between time steps, see Sect. 3, it is more practical to choose the neighborhood sufficiently large so that the system becomes overdetermined [4], therefore $|N_i| > M$. This *point cloud management* is discussed in more detail in [11] and also touched on in [15].

The overdetermined system (3) is then solved in a least squares sense, whereas an additional radial weighting function

$$w_i(r(\mathbf{x}_i, \mathbf{x}_j)) = \begin{cases} w_i \geq 0, & \text{if } r(\mathbf{x}_i, \mathbf{x}_j) < 1 \\ 0 & \text{otherwise} \end{cases} \tag{7}$$

is introduced. Because values at points that are closer to the central point $\mathbf{x}_i$ have a greater impact on the derivative at $\mathbf{x}_i$, the weighting function should be a non-increasing function [4]. The distance function being used is

$$r(\mathbf{x}_i, \mathbf{x}_j) = 2\frac{\|\mathbf{x}_i - \mathbf{x}_j\|}{h(\mathbf{x}_i) + h(\mathbf{x}_j)} . \tag{8}$$

This links the size of the support of w_i to the local smoothing length $h(\mathbf{x}_i)$. With

$$W_i = \begin{pmatrix} w_i(r(\mathbf{x}_i, \mathbf{x}_{J_1})) & & \\ & \ddots & \\ & & w_i(r(\mathbf{x}_i, \mathbf{x}_{J_{|N_i|}})) \end{pmatrix} , \tag{9}$$

$J_k \in N_i,\ k = 1, \ldots, |N_i|$, we can formulate the conditions for $\mathfrak{s}_i$ in a least squares sense:

$$\frac{1}{2}\left(\mathfrak{s}_i^*\right)^T W_i^{-2} \mathfrak{s}_i^* \to \min \tag{10}$$

under the conditions (3). This minimization problem can be solved using Lagrangian Multipliers by solving the linear system

$$K_i^T W_i^2 K_i \boldsymbol{\lambda}_i = -\mathbf{b}^* , \tag{11}$$

yielding a solution for $\mathfrak{s}_i^*$:

$$\mathfrak{s}_i^* = -W_i^2 K_i \boldsymbol{\lambda}_i = \left(W_i^2 K_i\right)\left(K_i^T W_i^2 K_i\right)^{-1} \mathbf{b}^*. \tag{12}$$

See [10], Theorem 5.1, for details. The stencils $\mathfrak{s}_i$ at all points of the pointcloud together yield a linear system $A\mathbf{u} = \mathbf{f}$. Solving this linear system gives the discrete approximation to the solution of the continuous problem (1).

3 The Finite Pointset Method (FPM)

Our work on linear solvers for GFDM is mainly motivated by a method called Finite Pointset Method (FPM) [2, 16, 17]. FPM is a Lagrangian method using GFDM in order to solve the incompressible Navier-Stokes equations using a pressure correction approach. The most important feature of FPM is that it is a Meshfree Method. It uses a point cloud in order to discretize the domain. All differential operators are discretized using this point cloud rather than using a fixed mesh or fixed cells. To this end, the techniques described in Sect. 2 are used. The fact that there is no fixed mesh involved makes methods like FPM very flexible when it comes to moving geometries or free surfaces: The points at the free surfaces or the moving parts of the geometry can be moved without a need to re-mesh the domain. FPM in particular also allows to introduce or delete points as needed.

In every time step, there are five linear systems that need to be solved: three Poisson-like pressure systems, one system for the velocity field and one system for the temperature.

In practice, the temperature equation as well as the vectorial velocity equation can be solved easily using a one-level method like BiCGStab(2) [12], which is a more stabilized version of the BiCGStab algorithm [19] that we will use for comparison later in this paper.

The Poisson-like pressure systems however pose problems to one-level methods when the point cloud becomes very large or when the geometry is stretched, as we will see in the Sects. 4 and 6.

3.1 Discretization of the Boundary

In Meshfree Methods, imposing boundary conditions is often quite involved. This also holds for FPM (cf. [8] and [15]) and has implications for the linear systems as well. Like the interior of the domain, the boundary is represented by a discrete set of points placed on the boundary. Boundary conditions are imposed by either setting up the stencil corresponding to the desired boundary operator (e.g. for Neumann boundary conditions, see [15] and also [8]) at the boundary points or—in case of Dirichlet boundary conditions—by setting the central stencil entry to 1 and the respective entry on the right hand side to the desired value. Note that therefore the central stencil values are

- $\mathcal{O}(1)$ at Dirichlet boundaries,
- $\mathcal{O}(h^{-1})$ at Neumann boundaries and
- $\mathcal{O}(h^{-2})$ in the interior.

4 Properties of the Linear Systems Arising in GFDM

This section will deal with the linear systems $A_h \mathbf{u}_h = \mathbf{f}_h$ arising from GFDM discretizations on point clouds as introduced in Sect. 2. The observations we make here are true in particular for the linear systems that need to be solved in FPM. It is important to be aware of those properties because they determine what linear solvers can be employed to solve the linear systems and how efficient they will be.

Throughout this section, we will use some terms and refer to some conditions that are well-known in the context of AMG and Multigrid Methods in general. At the beginning of Sect. 5 we give a very brief overview of the Multigrid algorithm. Interested readers are referred to [14] and [18] for an extensive explanation of the method.

Table 1 Properties of FDM and GFDM matrices

	Finite differences	Generalized finite differences
Symmetry	+	−
M-Matrix	+	−
Essentially irreducible	+	(+)
Sparsity	+	(+)
Elimination of boundary conditions	+	−

The more complex the model to be simulated becomes or the more accurate the results of the simulations have to be, the more points we need in order to discretize the simulation domain. Therefore, the number of degrees of freedom in the linear systems that we need to solve increases. Solving the linear systems becomes a bottleneck as most linear solvers do not scale linearly with the number of degrees of freedom.

Multigrid Methods on the other hand scale linearly in the number of degrees of freedom, their number of operations is $\mathcal{O}(n \log \varepsilon)$, i.e. their iteration number is $\mathcal{O}(\log \varepsilon)$, where ε is the desired relative residual reduction. Therefore, they are very well suited especially for large problems. However, not every linear system is suited to be solved using Multigrid Methods. There are certain properties of the corresponding matrix A that have to be fulfilled in order to theoretically guarantee convergence. Some other properties influence the efficiency of different linear solvers. Table 1 shows some of these properties and compares matrices arising from classical FDM to those from GFDM with respect to these properties.

Note that theoretically verified AMG methods only exist for symmetric, positive-definite M-Matrices (using Ruge-Stüben coarsening, [14]) or also non-symmetric M-Matrices (using a special aggregative coarsening strategy, [6]).

4.1 Symmetry and M-Matrix Property

In GFDM on point clouds, the stencil at every point is constructed based on local information only. Equation (12) shows that the stencil entries at $\mathbf{x}_i$ depend on the distances and the positions of the points in the neighborhood N_i only. Since any two distinct neighborhoods N_i and N_j will be different from each other, any two stencils will also be different from each other.

For this reason, there is no guarantee that $a_{ij} = a_{ji}$ in the assembled matrix A. In fact, it is also possible to have $a_{ij} \neq 0$ but $a_{ji} = 0$, which makes the matrix structurally non-symmetric. Not only does the Least Squares approach yield non-symmetric matrices, but it also does not guarantee non-positive stencil entries away from the central point. In consequence, the resulting matrix is not necessarily an M-Matrix.

4.2 Essentially Irreducible

When working with a fixed mesh with fixed neighborhood relationships like in standard FDM, a proper meshing of the domain will make sure that every point of the mesh is connected to at least one Dirichlet point, assuming a Dirichlet boundary is available. Point $\mathbf{x}_i$ is *connected* to $\mathbf{x}_k$ if there exists a path between the two points in the *adjacency graph* corresponding to the mesh. Then, in the corresponding matrix A every row will be connected to at least one Dirichlet row, which makes A an *essentially irreducible* matrix. See [11] for more thorough definitions of those terms.

This idea also carries over to GFDM on point clouds: Two points $\mathbf{x}_i$ and $\mathbf{x}_j$ are considered adjacent, if $a_{ij} \neq 0$ or $a_{ji} \neq 0$ in the corresponding GFDM matrix A. For a connected domain Ω and a sufficiently fine point cloud, GFDM will also yield essentially irreducible matrices. However, one of the main advantages of methods like FPM is that the point cloud is moving in every time step. Now consider a problem with free surfaces, for example a car driving through a body of water. In the course of the simulation, drops of water leave the main body of water and therefore the simulation domain is not connected any more. The full matrix A is then not essentially irreducible any more, however the block of A corresponding to the main body of water still is. One-level solvers are not sensitive to the full matrix A not being essentially irreducible, however Multigrid Methods are. We therefore employ a parallel algorithm to find independent submatrices in A. The submatrices corresponding to the drops of water may not be essentially irreducible any more, so we employ one-level or direct solvers to solve them. For the submatrix corresponding to the main body of water, we can employ Multigrid Methods as it is still essentially irreducible.

In this paper, we will limit ourselves to matrices that are essentially irreducible.

4.3 Sparsity

Since checking for and ensuring geometric conditions on the point distribution in every neighborhood to guarantee a solution of the linear system (3) would be rather involved (cf. Sect. 2) most methods including FPM use more neighboring points to construct the stencils than would be necessary in the optimal case. Note that for FPM there are methods to construct *minimal* stencils [11], which are not applicable to every situation, though. For the Poisson problem, which we need to solve in FPM, there exist Finite Difference stencils of second order with five (2D) and seven (3D) non-zero entries per point. In FPM on the other hand, at least 20 (2D) or 40 (3D) neighbors are used [15]. In classical AMG, this will lead to even less sparse coarse level operators because the Galerkin product RA_hP is used to construct the coarse level operator [14]. Less sparsity means more computational work for the linear solver. In the Multigrid framework, this is especially important for the *smoother* (cf. [14]). It is therefore important to achieve a good balance between having enough neighbors to ensure that (3) can be solved, but at the same time keep the number of neighbors as low as possible to minimize the computational effort needed in the linear solver. To this end, a nice property of the method we are proposing in Sect. 5 is that it allows to control the number of neighbors on the coarse levels to some extend.

4.4 Elimination of Boundary Conditions

In matrices arising from FDM, Neumann and Dirichlet boundary conditions are usually eliminated from the matrix and only affect the right hand side of the linear system. Due to the nature of GFDM on point clouds, this is not as easy with the matrices we are looking at. Let A_{II} be the submatrix of couplings from interior to interior points, A_{IB} and A_{BI} the submatrices of couplings from interior to boundary and boundary to interior points respectively, and finally let A_{BB} be the submatrix of couplings from boundary to boundary points. With a suitable renumbering of the points we then have to solve

$$\begin{pmatrix} A_{II} & A_{IB} \\ A_{BI} & A_{BB} \end{pmatrix} \mathbf{u} = \begin{pmatrix} \mathbf{f}_I \\ \mathbf{f}_B \end{pmatrix} . \tag{13}$$

Note that $(A_{IB})^T \neq A_{BI}$. Also note that, independently of the boundary conditions, the entries in A_{II} are of order $\mathcal{O}(h^{-2})$, $A_{II} \in \mathcal{O}(h^{-2})$ for short. On the other hand, the scaling of the entries in A_{BB} depends on the boundary conditions used. For pure Dirichlet boundary conditions for example, $A_{BB} \in \mathcal{O}(1)$. Neumann conditions on every boundary would lead to $A_{BB} \in \mathcal{O}(h^{-1})$. This obviously has a very negative impact on the range of eigenvalues of the linear system (see Sect. 4.5). In addition to that, the different scales of the interior part versus the boundary part of the matrix start to mix on the first coarse level when using AMG, because the coarse level operator is determined by using the Galerkin product $A_H = RA_hP$. The multigrid approach presented in Sect. 5 does not suffer from this situation as it constructs the coarse grid operator based on a coarser point cloud rather than on the fine level matrix. It is impractical to use a Schur complement ansatz and solve

$$\begin{pmatrix} A_{II} - A_{IB}A_{BB}^{-1}A_{BI} & 0 \\ A_{BI} & A_{BB} \end{pmatrix} \mathbf{u} = \begin{pmatrix} \mathbf{f}_I - A_{IB}A_{BB}^{-1}\mathbf{f}_B \\ \mathbf{f}_B \end{pmatrix} , \tag{14}$$

which would decouple the interior part from the boundary part, because for most applications, A_{BB} is too big to be inverted directly and it does not have any well defined properties that would allow a robust application of any iterative method to approximate the inverse. Even if A_{BB}^{-1} is available or easy to compute, then it is also unclear what properties $A_{II} - A_{IB}A_{BB}^{-1}A_{BI}$ has, which would then make it difficult to solve

$$\left(A_{II} - A_{IB}A_{BB}^{-1}A_{BI}\right)\mathbf{u} = \left(\mathbf{f}_I - A_{IB}A_{BB}^{-1}\mathbf{f}_B\right) \tag{15}$$

in the next step.

Scaling the original matrix A row-wise so that $a_{ii} = 1$ for every i, i.e. using

$$\hat{A} = \begin{pmatrix} a_{11} & & \\ & \ddots & \\ & & a_{nn} \end{pmatrix}^{-1} A \tag{16}$$

instead of A, yields a system with a smaller range of eigenvalues (see Fig. 1) and improves the convergence of BiCGStab (see Figs. 5, 7 and 8). In the context of AMG, a row-wise scaling is usually not advisable as it makes a symmetric matrix non-symmetric and destroys the symmetry of the couplings a_{ij} and a_{ji}, which is used in the coarsening process. In our case, however, we have a non-symmetric matrix to begin with and the notion of strong couplings in our matrix is non-symmetric from the beginning. The scaling (16) therefore does not sacrifice any properties of the matrix.

4.5 Eigenvalues

The eigenvalues of A are important to us, as they influence the convergence of most linear solvers. BiCGStab for example is known to show convergence rates that depend on the absolute range of eigenvalues, although this has not been proven theoretically. Note that when we talk about "smallest" and "largest" eigenvalues, we mean "smallest" or "largest" by absolute value, as the matrices we are dealing with are generally non-symmetric (see Sect. 4.1). For the purpose of this analysis we will restrict ourselves to the Poisson equation

$$\begin{cases} \Delta u & = f \text{ in } \Omega \\ u & = g \text{ in } \partial\Omega \end{cases} \tag{17}$$

with $\Omega = [0, L_1] \times [0, L_2]$. The point clouds we are working with are generated by a uniform distribution across Ω, but we make sure that the minimum distance between any two points is $0.1h$ and we choose $\beta = 1$ to define our neighborhood radius (see Definition 1).

Refining the Point Cloud Since FPM is often applied to problems with fairly detailed geometries, a sufficiently dense point cloud is required to correctly discretize the domain. Consider the Poisson equation (17) on the unit square, i.e. $L_1 = L_2 = 1$, with a varying smoothing length h. Figure 1 shows the smallest and largest eigenvalues of the matrix A discretizing the Laplacian in (17) with and without the normalization introduced in Sect. 4.4. While in both cases the range of eigenvalues increases when the point cloud becomes finer, the original matrices show the larger range with four to six orders of magnitude and the normalized matrices show the smaller range with two to four orders. We therefore expect BiCGStab to converge faster when using the normalized matrices and the results in Sect. 6 confirm that.

Stretching the Domain We are also interested in the eigenvalues for stretched domains. Here, we keep the smoothing length $h = 0.03$ fixed and focus on the normalized matrix, as this is the matrix with the smaller range of eigenvalues and we therefore expect it to yield better convergence rates. Our experiments for a domain stretched by a factor of 3.5, i.e. for $\Omega = [0, 3.5]^2$, yield a minimal eigenvalue that is about an order lower than for the original problem on $\Omega = [0, 1]^2$, see Fig. 2. In Sect. 6.3, we will come back to this observation.

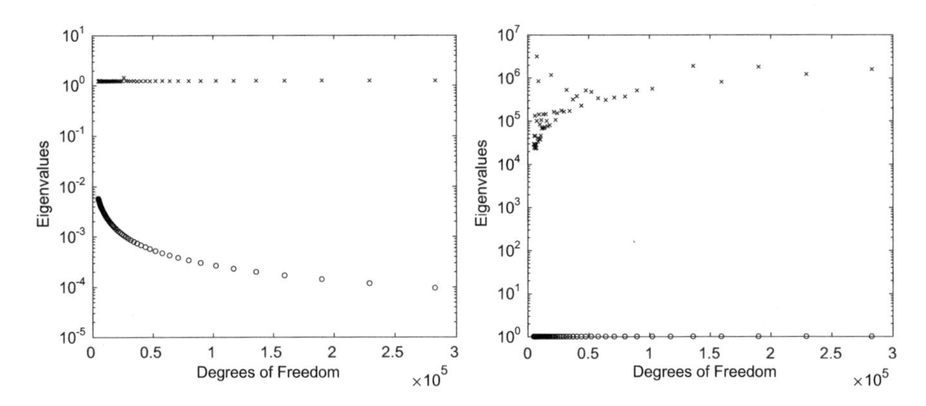

Fig. 1 Smallest and largest eigenvalues (by absolute value) of the normalized matrix $\hat{A}$ (left) and the original matrix A (right) when refining the point cloud

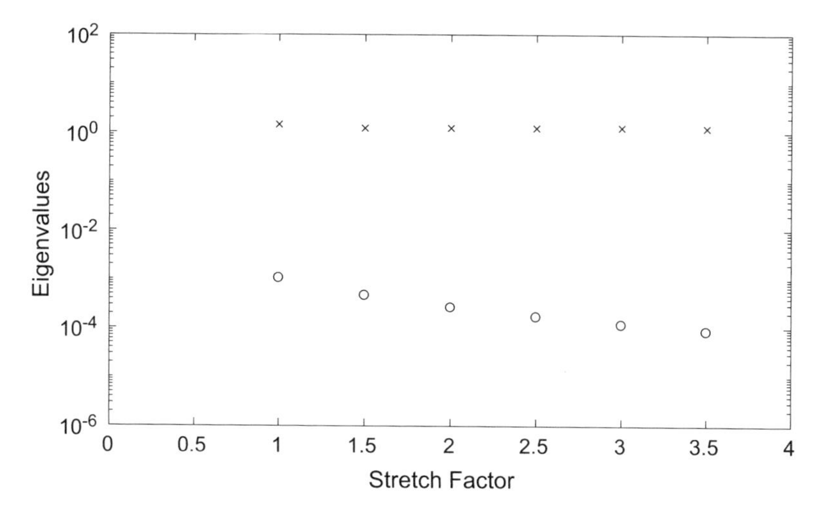

Fig. 2 Smallest and largest eigenvalues (by absolute value) of the normalized matrix $\hat{A}$ when stretching the domain in both directions

5 A Geometric Multicloud Approach

This section introduces a new iterative solver that is based on a hierarchy of point clouds and is related to Multigrid Methods.

The basic idea of an iterative two-level Geometric Multigrid Method is to construct a smaller linear system based on a *coarse grid* which is used to compute a correction by *restricting* the residual from the fine grid to the coarse grid. This correction is then *interpolated*, resulting in a correction for the original system. In order for this *coarse grid correction* to work, a *smoother* needs to be employed on the fine grid. Of course we can apply this idea recursively: Instead of solving the smaller liner system directly, we can construct an even smaller linear system using the same ideas. This process continues until the linear system is small enough to be efficiently solved by a direct solver. In general, Multigrid Methods follow the following scheme:

1. Smooth the error by applying ν_1 iterations of the *smoothing operator* M to the current approximation $\mathbf{u}$:

$$\tilde{\mathbf{u}} = M^{\nu_1}\mathbf{u} \tag{18}$$

2. Compute and restrict the residual to the coarse grid:

$$\mathbf{R} = R\left(\mathbf{f} - A_h\tilde{\mathbf{u}}\right) = R\mathbf{r} \tag{19}$$

3. Solve the (smaller) coarse grid equation either recursively or directly:

$$\mathbf{E} = A_H^{-1}\mathbf{R} \tag{20}$$

4. Interpolate the computed correction back to the fine grid:

$$\mathbf{e} = P\mathbf{E} \tag{21}$$

5. Apply the correction and another ν_2 iterations of the relaxation:

$$\mathbf{u}_{\text{new}} = M^{\nu_2}\left(\tilde{\mathbf{u}} + \mathbf{e}\right) \tag{22}$$

6. Continue with step 1 until the approximation $\mathbf{u}_{\text{new}}$ fulfills a specified termination criterion.

For details on Multigrid Methods, see [18].

Algebraic Multigrid solvers (AMG) remove the need of using a coarser grid to construct the smaller linear system. Instead, they use the size of the entries in the matrix itself. They can be very efficient solvers for many problems and it turns out that they also work very well for the systems arising in FPM, if the matrices are pre-processed correctly and the different components of AMG are tuned properly, see [5]. Because in AMG there are no "grids" involved, most authors use the term "levels" instead. In the following, we will also use the term "levels" as we are working in the field of Meshfree Methods which also implies that there are no "grids" involved in the classical sense. See [14] for details on AMG.

The simple structure of a point cloud with no inherent neighboring relationships however presents itself to using a geometric idea of creating coarse levels, namely by using coarser point clouds, rather than using the matrix entries. To generate coarse point clouds, there are two options:

- Generate multiple, independent point clouds with different smoothing lengths h or
- start with an existing, fine, point cloud and select a subset of the points to form a coarse point cloud.

Because in the context of Multigrid Methods it is advantageous to have the coarse level degrees of freedom be a subset of the fine level degrees of freedom, our method uses the second approach. Note that this is similar to the idea of a *Tree Partition of Unity Method* presented in [9], but works the opposite way: Instead of starting with a coarse point cloud and *refining* the point cloud successively, we are starting off with a fine point cloud that we will *coarsen*. This coarsening can be done based on geometric information only and no information on the stencils computed on the fine point cloud are needed. Therefore, the information on which we are basing our coarsening, namely the distance between two points, is symmetric, as opposed to the size of the matrix entries a_{ij} which are not necessarily symmetric in our case.

5.1 Coarsening Strategies Using Geometric Information

The coarsening strategy of the method we are proposing is based on the assumption that the solution of the linear system has similar values at points that are geometrically close to each other, which is the case especially in the pressure systems that need to be solved in FPM as long as the density ϱ is constant or at least does not have any jumps.

For a given fine point cloud $\mathcal{P}_h$ we start by choosing one point $\mathbf{x}_i$ as a *coarse level point*. Then all points that are within a distance of ch $(c < 1)$ of $\mathbf{x}_i$ become *fine level points*, meaning that they will not be present in the coarse point cloud $\mathcal{P}_H$, see Fig. 3. Note that we do not need to recompute any distances here, as they are already known from the operator setup, see Sect. 2. We repeat this procedure with every point that has not been assigned a label *coarse* or *fine*, yet. All points that have been assigned the *coarse* label form our coarse point cloud. Figure 4 shows how this procedure affects the neighborhood of a central point $\mathbf{x}_i$. The resulting point cloud does depend on the order in which we assign the *coarse* label to the points in the fine point cloud. A discussion on the impact of this order is beyond the scope of this paper though and will be discussed in future work.

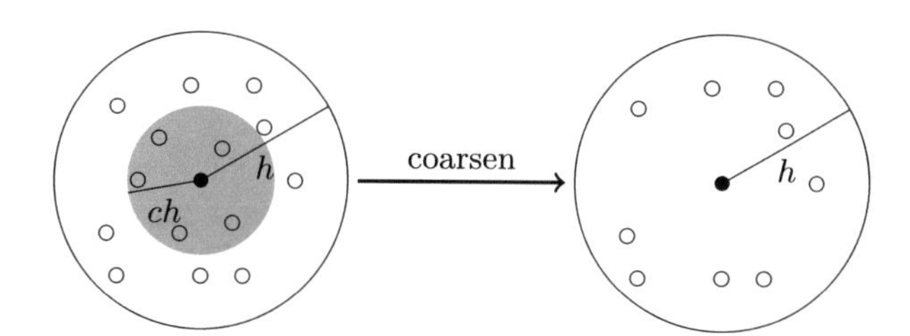

Fig. 3 Local view of the geometric coarsening strategy: points that are close to the central point (black) are not used to form the coarse point cloud

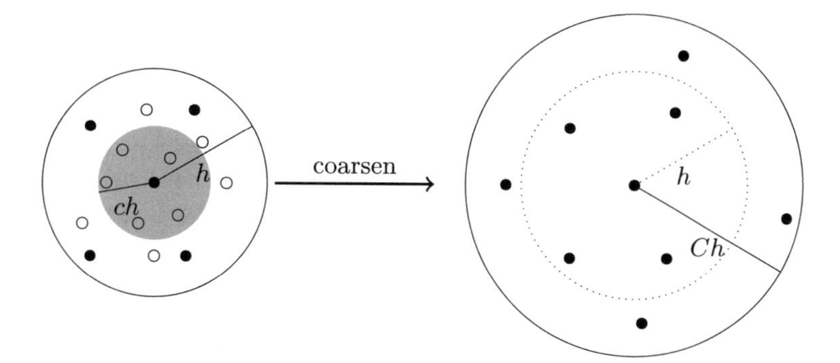

Fig. 4 After applying the coarsening strategy at every point in the fine level point cloud, the points labeled *coarse* (black) form the coarse level point cloud

However, because we want to use the coarse point cloud to construct the coarse operator—see Sect. 5.2—we need to make sure that the coarse point cloud is a proper discretization of the original problem. In particular, this means that there need to be boundary points in the coarse point cloud as well. The easiest way to make sure that not all of the boundary points are marked as fine level points is to start by applying the coarsening algorithm at the boundary points and only start marking interior points as coarse level points when all of the boundary points have been marked as fine or coarse level points. Figure 4 also shows that we need to reconsider what the coarse level neighborhood of the central point $\mathbf{x}_i$ is. Because we do not take the points near the central point to the coarser level, there are less neighbors within the old neighborhood radius h. If we would define the neighborhood on the coarse level with the same radius h that we use on the fine level, the situation depicted in Fig. 4 would cause issues. The way the coarse level operator is constructed (see Sect. 5.2) makes it mandatory for the coarse point cloud to allow the construction of an operator by the same method that was used to construct the fine level operator. In the situation depicted here, however, the central point $\mathbf{x}_i$ would not have a neighbor to the right, which would make the construction of a second order stencil impossible. Recall from Sect. 2 and also from Sect. 4.3 that

- making sure the points are distributed in such a way that the construction of a second order stencil is always possible is not straightforward and would also be too expensive in terms of run time and that
- FPM assumes that with 20 (2D) or 40 (3D) neighbors there is usually enough neighbors to construct second order stencils.

We stick with this assumption and therefore need to make sure that the number of neighbors in the coarse point cloud is about the same as in the fine point cloud. By using $H = Ch$ $(C > 1)$ as our new neighborhood radius and assuming that the points are uniformly distributed in the domain and assuming that the order in which we assign the *coarse* label does not affect this uniformness,

$$C = \sqrt{\frac{n_{\text{coarse}} + n_{\text{fine}}}{n_{\text{coarse}}}} \tag{23}$$

is sufficient to keep the average number of neighbors constant. Here, n_{coarse} and n_{fine} is the global number of coarse level and fine level points, respectively. Note that these are two disjoint sets and their union form the fine level point cloud. The parameter $c < 1$ can be used to determine the coarsening rate of the method. A value close to one means that more points in the neighborhood of a coarse level point will become fine level points, therefore there will be less points on the coarse level. A small value of c on the other hand will mean that less points will be declared fine level points.

5.2 Coarse Grid Operator

After having coarsened the point cloud, we can now construct a coarse grid operator on this point cloud. Instead of using a Galerkin operator like in AMG, we construct the coarse grid operator by applying the same scheme from Sect. 2 that was used on the fine point cloud to the coarse point cloud. This gives us a coarse grid operator discretizing the same continuous problem as the original one, but with a coarser mesh size and therefore a smaller matrix. As in all Multigrid Methods, we can either solve the coarse grid problem directly or use the same idea of coarsening recursively.

5.3 Restriction

Restriction is done based on the spatial distance between two points. The idea is that every coarse level point $\mathbf{x}_i$ has a coarse level residual $\mathbf{R}_i$ that is a weighted sum of the fine level residual $\mathbf{r}_i$ at the coarse point itself and the fine level residual at the fine level points in its neighborhood. Here, the weights $1/4$ and $3/4$ are loosely based on the idea of *full weighting* in Geometric Multigrid Methods [18]:

$$\mathbf{R}_i = \frac{1}{4}\mathbf{r}_i + \frac{3}{4} \sum_{j \in N_i^F} \alpha_{ij} \mathbf{r}_j \ , \tag{24}$$

where N_i^F is the subset of neighbors of $\mathbf{x}_i$ that are fine level points and

$$\alpha_{ij} = \frac{\hat{d}_{ji}}{\sum_{l \in N_j^C} \hat{d}_{jl}} \quad \text{with} \quad \hat{d}_{ij} = ||\mathbf{x}_i - \mathbf{x}_j||^{-1} \tag{25}$$

are weights. Note that the definition of α_{ij} depends on the *coarse* level neighbors N_j^C of $\mathbf{x}_j$ rather than on the neighbors of $\mathbf{x}_i$. It ensures that all weights going out from a fine level point sum up to 1.

Special considerations are needed at points that are neighbors of boundary points. Because of the different scale at boundary points (cf. Sect. 3.1), we cannot restrict the residual to the boundary as easily. Therefore, while we do take boundary points into account when computing the denominator of the weights in (25), we reset all weights to 0 that refer to a boundary point, i.e. we have

$$\alpha_{ij} = \begin{cases} 0 & \text{if } \mathbf{x}_j \in \partial\Omega \\ \frac{\hat{d}_{ji}}{\sum_{l \in N_j^C} \hat{d}_{jl}} & \text{else} \end{cases} . \tag{26}$$

We then make sure that the sum of all weights coming in to a specific coarse level point $\mathbf{x}_i$ is one. Otherwise, the correction we get from our coarse grid correction has a wrong scale. This is achieved by simply row-scaling the restriction operator so that all row sums equal 1.

5.4 Interpolation

The setup of the interpolation operator is very similar to that of the restriction operator. For every fine level point $\mathbf{x}_k$ the interpolated correction from the coarse level is

$$\mathbf{e}_k = \sum_{i \in N_k^C} \gamma_{ik} \mathbf{E}_i \quad \text{with} \quad \gamma_{ik} = \frac{\hat{d}_{ik}}{\sum_{j \in N_k^C} \hat{d}_{jk}} , \tag{27}$$

where N_k^C denotes the subset of neighbors of $\mathbf{x}_k$ that are coarse level points and $\mathbf{E}_i$ is the correction computed on the coarse level at $\mathbf{x}_i$. As opposed to the restriction, the interpolation weights γ_{ik} depend on the neighborhood of $\mathbf{x}_k$ only. For coarse level points $\mathbf{x}_k$ we set $\mathbf{e}_k = \mathbf{E}_k$.

Similarly to what we do in the restriction, at the boundary we reset all weights that refer to interpolations from the boundary to the interior of the domain to 0. As we are also cutting off the restriction to the boundary, this means that we have decoupled the interior from the boundary in both transfer operators. Since we construct the coarse operator by computing new stencils on the coarse point cloud though, there are still couplings between the interior and the boundary in the coarse operator. Therefore the correction that is computed using the coarse operator will take effects of the boundary conditions into account. The results in Sect. 6 show that our approach works well in the case of Dirichlet boundary conditions.

6 Results

In this section, we will evaluate different linear solvers using the model problem (17) with

$$f(x, y) = -8\pi^2 \sin(2\pi x) \sin(2\pi y) , \tag{28}$$

$$g(x, y) = 0 . \tag{29}$$

The stopping criterion for all linear solvers is a relative reduction of the initial residual by ten orders of magnitude. For both AMG and our Multicloud approach we use Gauss-Seidel relaxation. In the Multicloud approach, we use $c = 0.4$

as the coarsening parameter which leads to an average coarsening rate of about $1/10$. The Multicloud approach is applied to the original matrix A, rather than a scaled version, in all experiments. SAMG, developed by Fraunhofer SCAI, serves as an AMG implementation for comparison. Like in Sect. 4.5, the point clouds are generated with a normal distribution, additionally making sure that the minimal distance between any two points is greater than $0.1h$. Again, we use $\beta = 1$ to determine the neighborhood radius.

6.1 *Refining the Point Cloud*

In the first experiment, we keep the domain $\Omega = [0, 1]^2$ fixed and use point clouds of different density, i.e. we vary the smoothing length h. The resulting point clouds have between 4876 and 913,520 points. Figure 5 shows how the number of iterations the one-level solver BiCGStab needs in order to converge increases with the problem size. This behavior is typical for a one-level method. Our Multicloud approach on the other hand is a Multigrid Method and benefits from using the hierarchy of multiple point clouds. As we would expect from a Multigrid Method, the approach needs an almost constant amount of iterations for all problems: The number of iterations only varies between 10 and 13 for the Multicloud approach but between 93 and 2,136 for BiCGStab on the original matrix. When normalizing the matrix, which decreases the range of eigenvalues as Sect. 4.5 shows, BiCGStab needs less

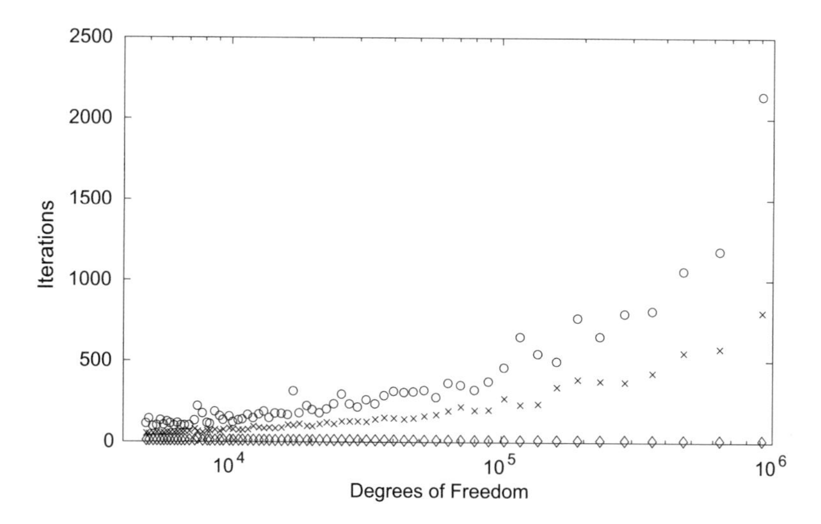

Fig. 5 Number of iterations needed to converge when decreasing h and therefore increasing the number of degrees of freedom. Open circle: BiCGStab on the original matrix; times symbol: BiCGStab on the normalized matrix; open diamond: Geometric Multicloud approach

iterations, however the number of iterations needed still increases significantly when refining the point cloud.

6.2 Comparing AMG to the Multicloud Approach

AMG methods are known to be very efficient solvers for sparse, elliptical linear systems, so they are a good benchmark to compare our method against. Note that in order to separate effects, we use both AMG and also our Multicloud approach without any acceleration, i.e. we do not use them as preconditioners for a Krylov method but in a stand-alone manner. AMG usually assumes that the given matrix is symmetric and positive definite, however it is known to work on matrices that do not fulfill these restrictions as well, at least within certain bounds. For this experiment, we use the original matrix A and our AMG method regards positive couplings as weak couplings (cf. [14]). The experiment in Fig. 6 shows that our method needs slightly less iterations compared to AMG, but both methods need an almost constant amount of iterations across the various refinement levels. We can therefore conclude that the Multicloud approach successfully makes use of the hierarchical structure of the problem by using coarse point clouds. If and how the advantage of less iterations carries over to a decrease in computational time will be the topic of future work.

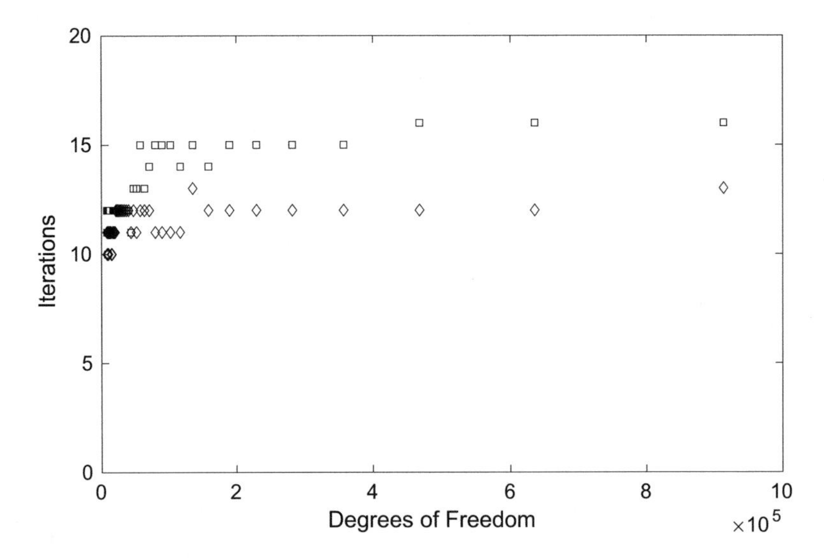

Fig. 6 Number of iterations needed to converge when changing h and therefore the number of degrees of freedom. Open square: AMG; open diamond: Geometric Multicloud approach

6.3 Stretching the Domain

Next, we consider the model problem (17) with a fixed smoothing length $h = 0.03$ on a varying domain, namely $\Omega = [0, L_1] \times [0, 1]$, $L_1 = 1, \ldots, 20$. This stretching of the domain does not have a significant influence on the number of iterations needed by BiCGStab or our Multicloud method, as is shown in Fig. 7. If we stretch the domain in both directions, i.e. consider the domain $\Omega = [0, L_1] \times [0, L_2]$, $L_1, L_2 = 1, \ldots, 5$, this observation changes. Figure 8 shows that in this case the number of iterations needed by the one-level BiCGStab method increases significantly when the domain gets larger. With 11 and 12 iterations the Multicloud approach again has an almost constant amount of iterations. The increase in iterations of the BiCGStab solver is more significant than the increased range in eigenvalues in Fig. 2 would indicate, which shows that the range of eigenvalues it not the only factor that determines the convergence rate of BiCGStab. Normalizing the matrix before applying BiCGStab helps again, but, as before, does not solve the problem of increasing iteration numbers.

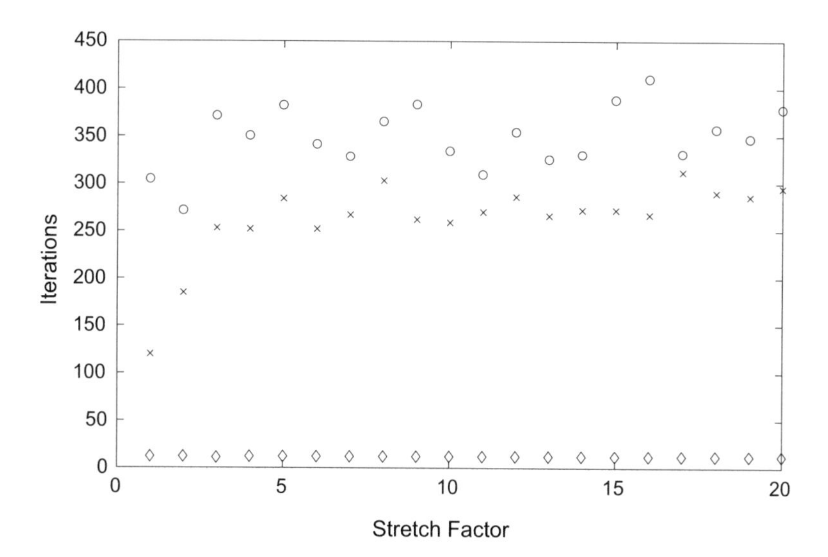

Fig. 7 Number of iterations needed to converge when stretching the domain in the x-direction only. Open circle: BiCGStab on the original matrix; times symbol: BiCGStab on the normalized matrix; open diamond: Geometric Multicloud approach. $h = 0.03$

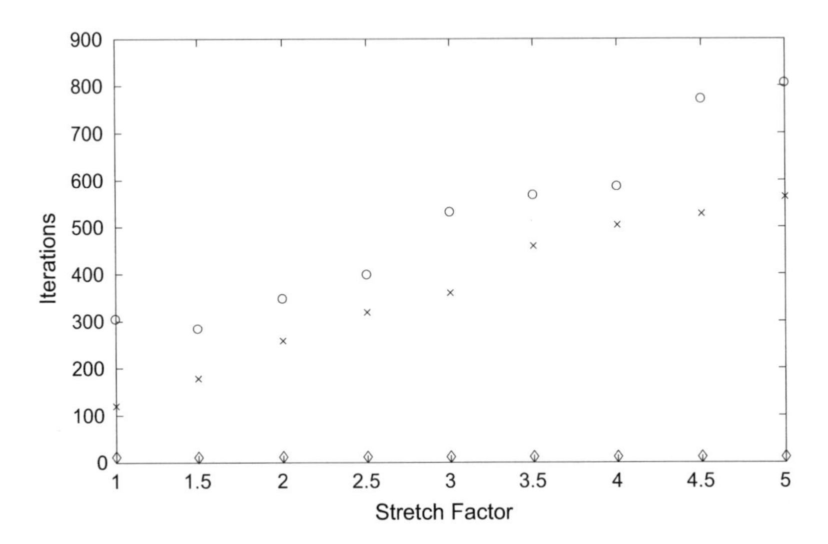

Fig. 8 Number of iterations needed to converge when stretching the domain in both x- and y-direction. Open circle: BiCGStab on the original matrix; times symbol: BiCGStab on the normalized matrix; open diamond: Geometric Multicloud approach. $h = 0.03$

7 Conclusions and Outlook

The present paper sheds some more light into the properties of the linear systems that can arise from GFDM and especially those arising in FPM. We saw that there are a lot of differences between the linear systems arising from GFDM compared to those arising from FDM. Those differences have implications regarding the linear solvers that can be used to solve those systems.

The Multicloud approach presented here has proven to yield convergence rates similar to those of AMG. Its convergence rate also does not decrease when refining the point cloud or stretching the geometry, which makes it a viable approach even for large problems. Its advantage over AMG is that it does not use any symmetry of the matrix because all its components are based on the underlying geometry only, which makes it more of a Geometric Multigrid Method.

On the other hand, we have seen that the one-level method BiCGStab shows the behavior of a non-constant number of iterations for a series of refined problems, like we know from FDM.

Future research will be dedicated to testing our Multicloud approach with boundary conditions other than Dirichlet boundary conditions. Neumann conditions might imply a need to re-introduce a coupling between the boundary and the interior in the transfer operators at those boundaries.

We will also examine the performance that can be achieved with the Multicloud approach in terms of computational time. For this, a tight integration with the discretization tool is needed, as the construction of the coarse operators is mainly done by that tool. A lot of information needed in the coarsening, the interpolation and restriction operators and also the construction of the coarse grid operator is already available, e.g. the neighborhood relationships and distances between points. There is no need to compute those again in the linear solver.

Another topic for future research is to examine what effect mixing components of AMG and the Multicloud approach has. If the Multicloud coarsening, interpolation and restriction are combined with a Galerkin coarse grid operator, this would give a method that does not use any symmetry in the matrix and at the same time does not need to compute any new stencils in order to set up the coarse grid operators. However, the decoupling of the boundary form the interior in the Multicloud transfer operators will be an issue here.

Finally, parallelization is an important aspect, too. The results with AMG in [5] already use a parallel AMG implementation, but there is no parallel implementation of the Multicloud approach, yet. In the light of parallelization, the performance difference between using the Galerkin operator versus constructing a coarse operator based on the original discretization approach will have to be revisited again as well.

References

1. Chen, J.S., Hillman, M., Chi, S.W.: Meshfree methods: progress made after 20 years. J. Eng. Mech. - ASCE **143**, 04017001 (2017)
2. Kuhnert, J.: Meshfree numerical scheme for time dependent problems in fluid and continuum mechanics. In Sudarshan, S. (Ed.) Advances in PDE Modeling and Computation, pp. 119–136. Ane Books, New Delhi (2014)
3. Lancaster, P., Salkauskas, K.: Surfaces generated by moving least squares methods. Math. Comput. **37**, 141–158 (1981)
4. Liszka, T., Duarte, C., Twordzydlo, W.: hp-Meshless cloud method. Comput. Methods Appl. Mech. Eng. **139**, 263–288 (1996)
5. Metsch, B., Nick, F., Kuhnert, J.: Algebraic multigrid for the finite pointset method. Comput. Vis. Sci. (2017, submitted)
6. Notay, Y.: An aggregation-based Algebraic Multigrid method. Electron. Trans. Numer. Anal. **37**, 123–146 (2010)
7. Perrone, N., Kao, R.: A general finite difference method for arbitrary meshes. Comput. Struct. **5**, 45–57 (1975)
8. Resńdiz-Flores, E.O., Kuhnert, J, Saucedo-Zendejo, F.R.: Application of a generalized finite difference method to mould filling process. Eur. J. Appl. Math. **29**, 1–20 (2017)
9. Schweitzer, M.A.: A Parallel Multilevel Partition of Unity Method for Elliptic Partial Differential Equations. University of Bonn, Bonn (2008)
10. Seibold, B.: M-Matrices in meshless finite difference methods. PhD thesis. University of Kaiserslautern (2006)
11. Seibold, B.: Minimal positive stencils in meshfree finite differnce methods for the poisson equation. Comput. Methods Appl. Mech. Eng. **198**, 592–601 (2008)
12. Sleijpen, G.L.G., Fokkema, D.R.: Bicgstab(l) for linear equations involving unsymmetric matrices with complex spectrum. Electron. Trans. Numer. Anal. **1**, 11–32 (1993)
13. Stellingwerf, R.F., Wingate, C.A.: Impact modeling with smooth particle hydrodynamics. Int. J. Impact Eng. **14**, 707–718 (1993)
14. Stüben, K.: An introduction to Algebraic Multigrid. Academic, Orlando (2000)

15. Suchde, P.: Conservation and accuracy in meshfree generalized finite difference methods. PhD thesis. University of Kaiserslautern (2017).
16. Tiwari, S., Kuhnert, J.: Finite Pointset Method Based on the Projection Method for Simulation of the Incompressible Navier-Stokes Equations. Springer, Berlin (2003)
17. Tiwari, S., Kuhnert, J.: Grid free method for solving Poisson equation. In Rao, G. (Ed.) Wavelet Analysis and Applications, pp. 151–166. New Age International Ltd. (2004)
18. Trottenberg, U., Oosterlee, C.W., Schüller, A.: Multigrid. Academic, New York (2000)
19. Van der Vorst, H.A.: Bi-CGSTAB: a fast and smoothly converging variant of Bi-CG for the solution of nonsymmetric linear systems. SIAM J. Sci. Stat. Comput. **13**, 631–644 (1992)

Additional Degrees of Parallelism Within the Adomian Decomposition Method

Andreas Schmitt, Martin Schreiber, and Michael Schäfer

Abstract The trend of future massively parallel computer architectures challenges the exploration of additional degrees of parallelism also in the time dimension when solving continuum mechanical partial differential equations. The Adomian decomposition method (ADM) is investigated to this respect in the present work. This is accomplished by comparison with the Runge-Kutta (RK) time integration and put in the context of the viscous Burgers equation.

Our studies show that both methods have similar restrictions regarding their maximal time step size. Increasing the order of the schemes leads to larger errors for the ADM compared to RK. However, we also discuss a parallelization within the ADM, reducing its runtime complexity from $O(n^2)$ to $O(n)$. This indicates the possibility to make it a viable competitor to RK, as fewer function evaluations have to be done in serial, if a high order method is desired. Additionally, creating ADM schemes of high-order is less complex than it is with RK.

Keywords Adomian decomposition · Burgers' equation · Runge-Kutta · Parallel in time

A. Schmitt (✉)
TU Darmstadt, Graduate School of Computational Engineering, Darmstadt, Germany

TU Darmstadt, Institute of Numerical Methods in Mechanical Engineering, Darmstadt, Germany
e-mail: aschmitt@gsc.tu-darmstadt.de; http://www.graduate-school-ce.de

M. Schreiber
University of Exeter, Mathematics/Computer Science, Exeter, UK

M. Schäfer
Graduate School of Computational Engineering, Technische Universität Darmstadt, Darmstadt, Germany

M. Schäfer et al. (eds.), *Recent Advances in Computational Engineering*, Lecture Notes in Computational Science and Engineering 124,
https://doi.org/10.1007/978-3-319-93891-2_7

1 Introduction

Simulations play an ever increasing role in science and industry. They are necessary, for instance, to simulate systems which cannot be studied experimentally such as planetary movements, to predict the weather of the next days, or to simulate molecular interactions for development of pharmaceutics. Very often, hard time constraints require to get simulation results within a reasonable time frame. Weather simulation is probably one of the most obvious examples. Additionally, driven by the stagnation in processor speed and the increasing degrees of on-chip parallelization over the last decade, exploring additional degrees of parallelism in the time domain is of steadily increasing interest. This is one of the main driving factors for parallel-in-time methods, see [23] for a review and overview.

Frequent choices of time stepping schemes from the family of the Runge-Kutta methods are the explicit, implicit, or implicit-explicit form [9, 20]. The application of these time stepping methods was already studied intensively. A less used and investigated method is the Adomian decomposition method (ADM) [6]. So far, the ADM was only considered as a time stepping method, but to our best knowledge no investigation was done regarding its properties of extracting additional degrees of parallelism. This motivates the present work in which we investigate the ADM in the context of parallelization and compare it to the explicit Runge-Kutta (RK) method for the viscous Burgers equation. Here, the focus lies on both the accuracy, which can be gained, and the time step restrictions.

1.1 Related Work

Adomian himself used the decomposition to calculate the exact solution for a specific initial velocity distribution [7]. Many authors approximated the exact solution by applying the Adomian decomposition using a truncated series instead of the originally used infinite series. Their work [8, 19, 21, 28, 29] shows that taking only a few terms of the series into account yields highly accurate results for the Burgers' equation. In [15, 22, 27] it is noted that truncating the series yields a small convergence radius, such that the maximal time step is bounded. Applying the ADM to an equation which is discretized in space reduces the radius even further [19].

The problem with the convergence radius was circumvented in [2] by using the ADM as a time stepping scheme with a stable time step size. Applying the ADM each time step makes spatial derivatives necessary for each interim solution. A reduction of complexity was used by approximating the interim solutions by an easily derivable series. With this method a good approximation of the exact solution was found. In [35] it is shown that the ADM is also able to yield very accurate results when the spatial domain is discretized and the ADM is used as a numerical time stepping scheme.

This work focuses on a comparison of the ADM with the RK method which was already done for other models. These comparisons were carried out numerically with a continuous spatial domain. For the Lorenz equation it was shown in [25] that the ADM with four terms of the series allows for larger time steps than the classical RK method and reaching the same order of accuracy. Another comparison based on the Lorenz equation compared the ADM with 15 terms of the series with Runge-Kutta-Verner schemes of fifth and sixth order which showed the ADM to be more accurate [33]. Using the ADM with four terms and comparing it to the classical RK method displayed errors of the same order for both methods tested on different linear and non-linear equations [32].

1.2 Contribution of This Work

Besides the standard stability and convergence requirements, a viable time stepping scheme in productive simulations has to fulfill two important properties: accurate solutions even for larger time step sizes and, related to larger time step sizes, a wall-clock time as small as possible. Considering the ADM, previous work has mainly focused on its accuracy. Our focus is on a reduction of the wall-clock time. This reduction can be achieved by (a) large time steps and by (b) speeding up the time stepping method by exploiting additional degrees of parallelism in the time dimension. Both parts are less investigated so far regarding the ADM as a time stepping method and this is the main focus of this work. By comparing the ADM to the RK method we investigate whether the discrete ADM is a viable method regarding real-world scenarios and if it is competitive to other existing time stepping methods.

In Sect. 2 we present a short introduction to the viscous Burgers equation. This is followed by the comparison of the Runge-Kutta and discrete Adomian decomposition methods in Sect. 3, where the discrete ADM is also described in more detail. The additional degrees of parallelism of the ADM are then discussed in Sect. 4, followed by results of the numerical studies in Sect. 5 and conclusions in Sect. 6.

2 Burgers' Equation

The Navier-Stokes equations (NSE) are the fundamental equations of many computational fluid dynamics (CFD) problems [34]. In their incompressible form they read

$$\frac{\partial \mathbf{u}}{\partial t} + (\mathbf{u} \cdot \nabla)\,\mathbf{u} = -\frac{1}{\rho}\nabla p + \nu \nabla^2 \mathbf{u} + \frac{1}{\rho}\mathbf{F}\,, \tag{1}$$

$$\nabla \cdot \mathbf{u} = 0\,,$$

where $\mathbf{u}$ denotes the velocity, p the pressure, ρ the density, ν the kinematic viscosity, and $\mathbf{F}$ the external forces.

The viscous Burgers equation, which was introduced by Bateman [11] and extensively studied by Burgers, e.g. [12], can be derived from (1). To gain Burgers' equation, we drop the internal sources $-\nabla p/\rho$ and external sources $\mathbf{F}/\rho$ leaving us with

$$\frac{\partial \mathbf{u}}{\partial t} + (\mathbf{u} \cdot \nabla)\,\mathbf{u} = \nu \nabla^2 \mathbf{u}\,.$$

This simplification of the NSE still contains the non-linearity, which is one of the terms of interest when it comes to developing numerical schemes for the NSE.

For a better presentation of our results, we use the one dimensional formulation

$$\frac{\partial u}{\partial t} + u\frac{\partial u}{\partial x} = \nu \frac{\partial^2 u}{\partial x^2}\,. \tag{2}$$

3 Comparison of Time Integration Methods

In this section, we compare the Runge-Kutta method to the discrete Adomian decomposition method and provide a more detailed description of both methods.

3.1 Runge-Kutta Method

In our comparison we use the well-known explicit Runge-Kutta method (RK) [30] and employ schemes of order $p = 1, \ldots, 4$. The RK method solving $\mathrm{d}u/\mathrm{d}t = f(t, u)$ reads

$$\begin{aligned} u_{n+1} &= u_n + \Delta t \sum_{i=1}^{s} b_i k_i, \\ k_1 &= f(t_n, u_n), \\ k_i &= f\left(t_n + c_i \Delta t, u_n + \Delta t \sum_{j=1}^{i-1} a_{ij} k_j\right), \qquad \text{for } i = 2, \ldots, s, \end{aligned}$$

where u_n denotes the approximation of $u(t_0 + n\Delta t)$, and the coefficients a, b, c are specific to the used s-stage RK scheme. These coefficients are usually arranged in a Butcher tableau [13]

$$\begin{array}{c|cccc} 0 & & & & \\ c_2 & a_{21} & & & \\ \vdots & \vdots & \ddots & & \\ c_s & a_{s1} & \cdots & a_{s,s-1} & \\ \hline & b_1 & \cdots & b_{s-1} & b_s \end{array} .$$

For an s-stage RK method of order p, $s \geq p$ holds and if $p \geq 5$, then $s > p$ [14]. From this follows, that as many or more function evaluations are necessary per time step as the order of the scheme is.

Since the RK method is of explicit nature it has a limited stability region. The time step is bounded by the condition that the physical propagation of information has to be equal to, or slower than the propagation of information on the discretized grid. This is reflected in the CFL condition [18]. With the Burgers equation the fastest propagation of information can either be by advection or diffusion making the maximal time step width proportional to $\Delta t_{\max} \propto \Delta x$ or $\Delta t_{\max} \propto \Delta x^2$, respectively.

3.2 Discrete Adomian Decomposition Method

Next, we describe the discrete Adomian decomposition method (DADM). For this purpose, we first provide an overview of the Adomian decomposition method, before we explain its discretized version. Finally, we apply the DADM to the Burgers equation.

Adomian Decomposition Method The basis of the DADM is the Adomian decomposition method which was developed by Adomian in the 1980s and 1990s, e.g. [6]. The main idea is to decompose the non-linearities of equations into a series of Adomian polynomials. With the ADM it is possible to calculate the analytical solution or, if this is not possible, to gain an approximation with a fast convergence to the actual solution. This is possible as no discretization (in case of the analytical version described here), linearization, or perturbation theory has to be applied to the non-linear term [3, 4, 6, 7].

Following the notation in [5] the method is in general applied to an equation $\mathcal{F}(u) = g$, where $\mathcal{F}$ is a (non-)linear operator. This operator can be decomposed in its linear and non-linear parts $\mathcal{L} + \mathcal{R}$ and $\mathcal{N}$, respectively. Here, $\mathcal{L}$ denotes an easily invertible operator of highest order and $\mathcal{R}$ comprises the remaining linear parts. Using this notation we get

$$\mathcal{L}u + \mathcal{R}u + \mathcal{N}(u) = g \; . \tag{3}$$

Let the inverse of $\mathcal{L}$ be $\mathcal{L}^{-1}$. Solving (3) for $\mathcal{L}u$ and applying the inverse yields

$$\mathcal{L}^{-1}\mathcal{L}u = \mathcal{L}^{-1}g - \mathcal{L}^{-1}\mathcal{R}u - \mathcal{L}^{-1}\mathcal{N}(u) \; . \tag{4}$$

The left hand side of this equation can be evaluated to $\mathcal{L}^{-1}\mathcal{L}u = u + C$, where C are the integration constants given either by initial or boundary conditions.

The idea of the ADM is now to expand the unknown into the series $u = \sum_{i=0}^{\infty} u_i$ and write the non-linearity as $\mathcal{N}(u) = \sum_{i=0}^{\infty} A_i$, where the A_i are the Adomian polynomials. With the series representation we can identify $u_0 = \mathcal{L}^{-1}g - C$. Now, we can write (4) as

$$\sum_{i=0}^{\infty} u_i = u_0 - \mathcal{L}^{-1}\mathcal{R}\sum_{i=0}^{\infty} u_i - \mathcal{L}^{-1}\sum_{i=0}^{\infty} A_i \; . \tag{5}$$

Therefore, we have the recursive relation

$$\begin{aligned} u_1 &= -\mathcal{L}^{-1}\mathcal{R}u_0 - \mathcal{L}^{-1}A_0 \\ u_2 &= -\mathcal{L}^{-1}\mathcal{R}u_1 - \mathcal{L}^{-1}A_1 \\ &\vdots \\ u_{i+1} &= -\mathcal{L}^{-1}\mathcal{R}u_i - \mathcal{L}^{-1}A_i \\ &\vdots \end{aligned} \tag{6}$$

for the u_i of the expansion series.

The Adomian polynomials can formally be written as

$$A_i(u_0, u_1, \ldots, u_i) = \frac{1}{i\,!}\left[\frac{\mathrm{d}^i}{\mathrm{d}\lambda^i}\mathcal{N}\left(\sum_{j=0}^{\infty}\lambda^j u_j\right)\right]_{\lambda=0} , \tag{7}$$

(see [1, 3, 4, 6, 16]). We notice that A_i is only dependent on $u_0, u_1, \ldots, u_i$ and other independent variables. This is important for the discussion of the parallelization in Sect. 4. In the end, the formulation of the A_i can be interpreted as a generalization of the Taylor series in the neighborhood of the function u_0

$$\mathcal{N}(u) = \sum_{i=0}^{\infty} A_i = \sum_{i=0}^{\infty}\frac{1}{i\,!}\left(u - u_0\right)^i \mathcal{N}^{(i)}\left(u_0\right) \; .$$

Discretization of the Adomian Decomposition Method Using the ADM numerically requires a truncation to a finite series. As stated in [15, 22, 27] this truncation leads to a small convergence radius of the ADM. In [35] this small convergence radius was circumvented by applying the ADM as a time stepping scheme iteratively on a discretized spatial domain. Since our ansatz is applying discretization in space and time as-well, we adapt the name discrete Adomian decomposition method.

The truncated version of the decomposition (5) to approximate the solution reads

$$\sum_{i=0}^{\infty} u_i \approx \sum_{i=0}^{p} u_i = u_0 - \mathcal{L}^{-1}\mathcal{R}\sum_{i=0}^{p-1} u_i - \mathcal{L}^{-1}\sum_{i=0}^{p-1} A_i \ ,$$

where p denotes the order of the approximation. The truncated version can be used as a time stepping scheme, by repeatedly applying the ADM with a stable time step size Δt. The starting point of each new time step is given by the result of the previous time step. We denote the approximation at time $t_n = t_0 + n\Delta t$ by $\hat{u}^n$. With this we can define the time stepping scheme of order p as

$$\hat{u}^{n+1} = \hat{u}^n + \sum_{i=1}^{p} u_i \ . \tag{8}$$

For one time step of this scheme p evaluations of the linear operator are necessary and the non-linearity has to be evaluated in the order of $O(p^2)$ times because of (7).

DADM Applied to Burgers' Equation We compare the DADM with the RK method applied to the Burgers equation. For the Burgers equation (2) we can identify $\mathcal{L}u = \partial u/\partial t$, $\mathcal{R}u = -\nu\partial^2 u/\partial x^2$ and $\mathcal{N}(u) = u\partial u/\partial x$. Using this and having (8) in mind the recursion (6) reads

$$\begin{aligned}
u_0 &= \hat{u}^n \\
u_1 &= -\frac{\Delta t}{1}(\mathcal{R}u_0 + A_0) \\
u_2 &= -\frac{\Delta t}{2}(\mathcal{R}u_1 + A_1) \\
&\ \vdots \\
u_p &= -\frac{\Delta t}{(p-1)}(\mathcal{R}u_{p-1} + A_{p-1}) \ .
\end{aligned} \tag{9}$$

Here, we utilized $\mathcal{L}^{-1} = \int_{t_n}^{t_{n+1}} \cdot \, \mathrm{d}t$ and $\hat{u}^n$ being independent on t. Equation (7) can be written as

$$A_i = \sum_{j=0}^{i} u_j \frac{\partial u_{i-j}}{\partial x} \tag{10}$$

in this context.

3.3 Comparison of DADM and RK

Both methods investigated in this comparison are explicit methods, which can be formulated with different orders of accuracy. We introduce the notation DADMp and RKp, where p denotes the order of the method. For the purpose of readability, we also introduce the notation $\mathcal{N}(x(u), y(u)) = x(u)\partial y(u)/\partial x$ for the non-linearity.

First Order We compare DADM1 with the explicit Euler method, which is equivalent with RK1. Writing out DADM1 yields

$$\hat{u}^{n+1} = \hat{u}^n + \sum_{i=1}^{1} u_i = \hat{u}^n - \Delta t \left(\mathcal{R}\hat{u}^n + \mathcal{N}\left(\hat{u}^n, \hat{u}^n\right)\right) ,$$

and using the same operator notation for RK1, we get

$$u^{n+1} = u^n - \Delta t \left(\mathcal{R}u^n + \mathcal{N}\left(u^n, u^n\right)\right) .$$

As we can see, both methods are exactly the same.

Second Order Writing $\hat{u}^{n+1}$ as a function of $\hat{u}^n$ for DADM2 yields

$$\begin{aligned}
\hat{u}^{n+1} &= \hat{u}^n + \sum_{i=1}^{2} u_i \\
&= \hat{u}^n - \Delta t \left(\mathcal{R}\hat{u}^n + \mathcal{N}\left(\hat{u}^n, \hat{u}^n\right)\right) \\
&\quad + \frac{\Delta t^2}{2} \left(\mathcal{R}^2\hat{u}^n + \mathcal{R}\mathcal{N}\left(\hat{u}^n, \hat{u}^n\right) + \mathcal{N}\left(\mathcal{R}\hat{u}^n + \mathcal{N}\left(\hat{u}^n, \hat{u}^n\right), \hat{u}^n\right)\right. \\
&\quad \left. + \mathcal{N}\left(\hat{u}^n, \mathcal{R}\hat{u}^n + \mathcal{N}\left(\hat{u}^n, \hat{u}^n\right)\right)\right) .
\end{aligned}$$

We can see, that this is the second order exact representation of the generalized Taylor series developed around the function $\hat{u}^n$, as expected with the ADM.

The RK2 method, which is the midpoint method, can also be written without intermediate stages

$$\begin{aligned}
u^{n+1} &= u^n - \Delta t \left(\mathcal{R}u^n + \mathcal{N}\left(u^n, u^n\right)\right) \\
&\quad + \frac{\Delta t^2}{2} \left(\mathcal{R}^2 u^n + \mathcal{R}\mathcal{N}\left(u^n, u^n\right) + \mathcal{N}\left(\mathcal{R}u^n + \mathcal{N}\left(u^n, u^n\right), u^n\right)\right. \\
&\quad \left. + \mathcal{N}\left(u^n, \mathcal{R}u^n + \mathcal{N}\left(u^n, u^n\right)\right)\right)
\end{aligned}$$

$$+\frac{\Delta t^3}{4}\left(\mathcal{N}\left(\mathcal{N}\left(u^n\right), \mathcal{R}u^n\right) + \mathcal{N}\left(\mathcal{R}u^n, \mathcal{N}\left(u^n, u^n\right)\right)\right.$$
$$\left. + \mathcal{N}\left(\mathcal{N}\left(u^n\right), \mathcal{N}\left(u^n\right)\right) + \mathcal{N}\left(\mathcal{R}u^n, \mathcal{R}u^n\right)\right) .$$

In this formulation we recognize all terms of DADM2. In addition, there are some terms of third order. These terms do not represent the third order terms of the generalized Taylor series and, therefore, impact the error but not the convergence order of the numerical scheme.

Higher Order Higher order methods show a comparable picture as the second order methods. The DADMp is, as intended, always accurate up to order p regarding the generalized Taylor series. RKp is based on the same terms as DADMp, but also has additional terms of orders higher than p.

Similar solutions are expected due to this from the numerical schemes. The remaining question is, whether the additional terms are beneficial for the stability (maximal stable time step width).

4 Degrees of Parallelism

The goal of this work is to investigate whether DADM is a viable method and competitive to other time steppers like RK. For this purpose, we have already compared both methods in Sect. 3. There, we found that both methods are very similar, but need different numbers of function evaluations. One could argue that the large amount of evaluations of the non-linear term in DADM makes this method less efficient than RK.

Taking a closer look at the data dependencies of the calculation of the Adomian polynomials, we can recognize (see also Sect. 3.2) that these are a sum over already calculated variables, sketched in Fig. 1. We can use this to parallelize the calculation of the terms in (7), respectively (10).

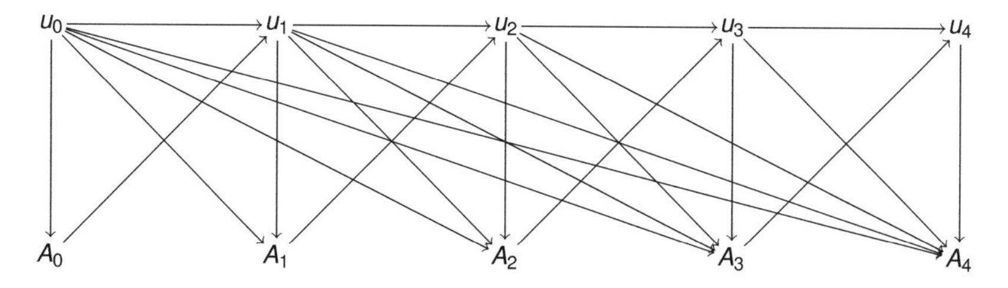

Fig. 1 Data dependencies indicated by the arrows for the first four Adomian polynomials A_i are shown to underline the fact that each polynomial is dependent on all previously calculated velocities u_i

Table 1 This table shows distribution of workload on multi-core system

	P_0	P_1	P_2	P_3	Calculations and dependencies
Bcast	u_0				
	A_0	A_0	A_0	A_0	$A_0 = (u_0 \cdot \nabla)u_0$
	u_1	u_1	u_1	u_1	$u_1 = -\Delta t(\nu\nabla^2 u_0 + A_0)$
Reduce	$u_0\nabla \cdot u_1$	$u_1\nabla \cdot u_0$			
	A_1	A_1	A_1	A_1	$A_1 = \sum_{i=0}^{1}(u_i \cdot \nabla)u_{1-i}$
	u_2	u_2	u_2	u_2	$u_2 = -\Delta t/2(\nu\nabla^2 u_1 + A_1)$
Reduce	$u_0\nabla \cdot u_2$	$u_1\nabla \cdot u_1$	$u_2\nabla \cdot u_0$		
	A_2	A_2	A_2	A_2	$A_2 = \sum_{i=0}^{2}(u_i \cdot \nabla)u_{2-i}$
	u_3	u_3	u_3	u_3	$u_3 = -\Delta t/3(\nu\nabla^2 u_2 + A_2)$

The left most column indicates the necessary communication. The columns headed by P_i denote the processors and show the values on the processor. In the last column the dependencies and calculations are given. Here, the calculation for the first three velocities is shown to demonstrate the idea

To our best knowledge, these additional degrees of parallelism were not studied in the literature so far. A strictly sequential evaluation of a p-stage DADM, leads to $O(p)$ function evaluations plus $\sum_{n=1}^{p} n = \frac{1}{2}p(p+1) = O(p^2)$ additional evaluations of the non-linear term following (9) and (10). However, each stage leads to additional independent terms which increases the degree of parallelism for higher stages over the sum. Assuming that the evaluations of each term in the stage are computationally more expensive compared to the reduction over the sum, this reduces the runtime complexity to $O(p)$. In Table 1 the workload distribution on a multi-core system is depicted. In addition to the distribution, we highlight the required data dependencies, hence the communication.

We would also like to point out that some terms of (10) can already be computed at an earlier stage, e.g. the term $u_1\nabla \cdot u_1$ belonging to A_2 can be evaluated at the time the terms for A_1 are evaluated. Hence, a formulation as a directed acyclic graph (DAG) scheduling problem could lead to further improvements regarding the wall-clock time and the required computational resources.

5 Numerical Studies

The numerical studies described in this section are used to expand the comparison of RKp and DADMp. We start with a description of the benchmark used for the studies. With this benchmark scenario we compare the maximal time step size and the convergence order of both schemes.

5.1 Spatial Discretization

As the interest of our study is the comparison of the time stepping methods, we apply spectral methods for the discretization of our spatial domain. With this we are able to minimize the errors induced by the spatial discretization of the equation.

The method of choice is the periodic Fourier basis, e.g. [20]. This basis allows to evaluate linear operators element-wise in spectral space, hence without loss of accuracy. Therefore, the error of this operation is in the order of machine precision, if the function can be represented exactly in spectral space with the given number of modes.

Non-linear operators are evaluated in a pseudo-spectral fashion [10, 24] in physical space to reduce the complexity of the calculation, which would be given with the necessary convolution of all spectral series. A standard anti-aliasing technique is used to filter the spurious modes created in physical space by the pseudo-spectral calculation, e.g. [31].

5.2 Benchmark Scenario

For our benchmark the spatial domain is initialized at $t = t_0$ with a one dimensional sinus wave as shown in Fig. 2. The space time domain, on which the benchmark is run, is $x, t \in [0, 1]^2$. Even with an initial distribution as simple as this, it would not be very easy to apply the ADM numerically without a spatial discretization.

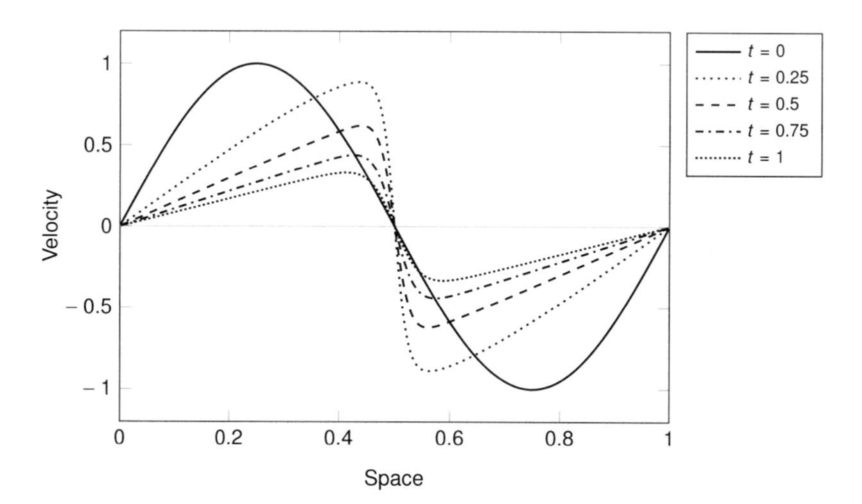

Fig. 2 Snapshots of the analytical solution to the viscous Burgers equation for the velocity over the spatial domain with $\nu = 0.01$

This is due to the convergence radius requiring multiple steps of the ADM and the intermediate solutions are not easily differentiable.

As the focus of the Adomian decomposition is the non-linearity, we use a viscosity of $\nu = 0.01$, which reduces the influence of the diffusion compared to the advection. With this choice the snapshots in Fig. 2 show that both parts have a visible influence on the solution. The spatial domain is discretized with a grid spacing of $\Delta x = 1/256$, which corresponds to $N = 170$ spectral modes. For the comparison of the convergence order, we run the schemes with time step widths of $\Delta t = \{1.25 \times 10^{-4}, 2.5 \times 10^{-4}, 5 \times 10^{-4}\}$. All errors are the difference between the numerical and analytical solution, which is given by the Hopf-Cole transformation [17, 26].

5.3 Stable Time Step Size Limitations

Since one of the possibilities to reduce the calculation time is to use large time step sizes, it is of interest to investigate the largest possible time step size with both methods. Therefore, we compare the maximal stable time step $\Delta t_{\max}$ for different orders of the DADM and RK method applied to the benchmark. The maximal stable time step for this investigation is the Δt with which a solution at the final time $t = 1$ was found without significant deviations from the analytical solution. The calculations were done with schemes of the orders $p = \{1, \ldots, 4\}$ for RK and with schemes of the order $p = \{1, \ldots, 15\}$ for DADM. Δt_{max} was determined with up to three significant digits.

In Fig. 3 the upper two curves show the results for the mentioned settings. Considering the schemes of order $p = \{1, \ldots, 4\}$, we can observe that both RK and DADM show very similar maximal stable time step sizes. For the even orders DADM allows time steps which are approximately 3% larger than those with RK. With an increasing order slightly larger time steps are possible. This trend continues for DADM for the other tested orders $p = \{5, \ldots, 15\}$.

The two bottom curves in Fig. 3 were calculated with $N = 341$ spectral modes. The smaller Δt_{max} is expected, because both schemes are of explicit nature and their time step size is bounded by the CFL condition. Increasing the number of spatial discretization points by a factor of two in this case leads to a reduction of the maximal stable time step width by a factor of four. From this we can deduce that the time step width is bounded by the diffusive information transport. In this case the relation $\Delta t_{max} \propto \Delta x^2$ holds.

From these results we can conclude that there seems to be no striking difference between RK and DADM considering Δt_{max}. Therefore, applying the parallelization described in Sect. 4 can make DADM competitive against RK for high order methods, as there is no disadvantage in time to solution regarding the time step width.

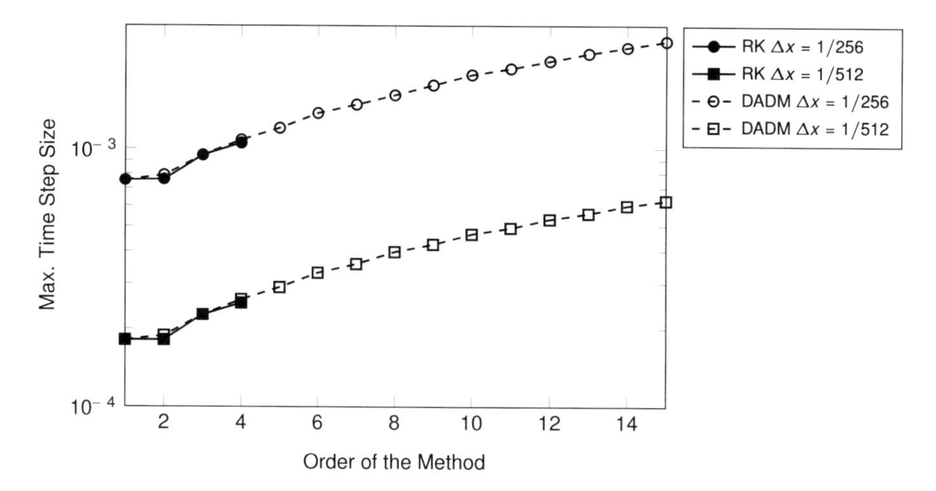

Fig. 3 Maximal time step Δt_{max} for converged calculations in time interval $t \in [0, 1]$ for RK (solid) and DADM (dashed) plotted over the order of the method

5.4 Convergence Order

In addition to the maximal time step size, we take a look at the errors produced by the schemes to investigate the assumption made in Sect. 3.3, where we stated that the additional higher order terms of RK will reduce the error compared to DADM. Here, we compare the errors for the orders $p = 1, \ldots, 3$ of both schemes and show that the schemes converge with the given order.

In Fig. 4 the maximal absolute error at the final time $t = 1$ is shown for the three different time step widths of the benchmark. As expected from the comparison in Sect. 3.3, the first order methods have exactly the same error and converge according to their order. The errors of DADM2 are slightly longer than those of RK2. The difference is a result of the additional terms of higher order which appear in RK2. Both schemes converge with second order accuracy. For the third order schemes we get a comparable picture. Although both methods reach the expected order, the DADM3 produces larger errors than the RK3. Taking all three orders into account, we can see that the error difference increases with increasing order.

From a serial point of view this and having comparable maximal time step sizes suggests to prefer the RK method over the DADM. With the discussed parallelization DADM is still viable. For smaller orders the error difference is not significant. In case of the higher orders increasing the order from p to $p + 1$ for the DADM needs still as many function evaluations as RK for $p \geq 5$ and even less for $p \geq 7$ [14].

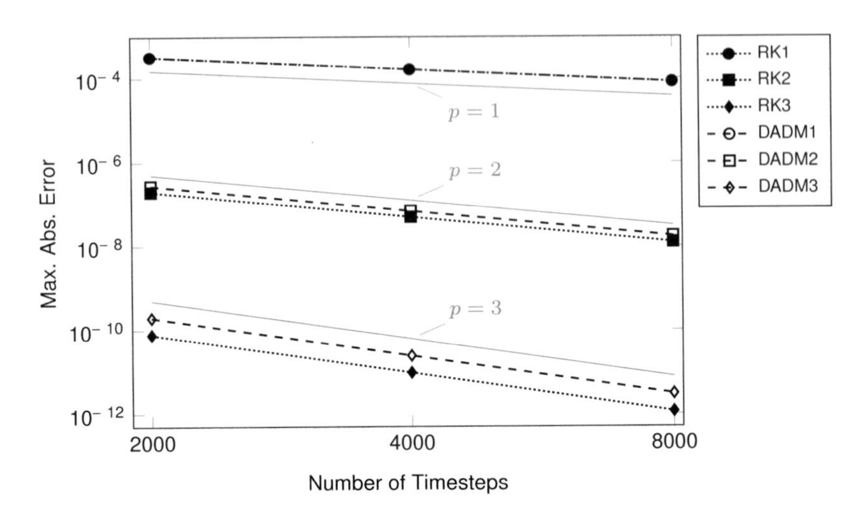

Fig. 4 Maximal absolute error at $t = 1$ plotted over the number of time steps for RK and DADM of order $p = 1, \ldots, 3$. Additionally, slopes of order $p = 1, \ldots, 3$ are shown to validate the order of the methods

6 Conclusion and Discussion

In this work, we have discussed how the Adomian decomposition method (ADM) can be applied as an explicit time stepping scheme. We were able to extract additional degrees of parallelism within this method to reduce the runtime complexity from $O(n^2)$ to $O(n)$.

This reduction makes the ADM a viable competitor to the Runge-Kutta (RK) method for high-order schemes. In this case the number of function evaluations which have to be computed in serial is smaller for the ADM than for the RK method. A comparison based on numerical studies has shown that both methods have comparable maximal time step widths. The larger errors of the ADM in the high orders can be circumvented by increasing the order of the ADM from p to $p + 1$. This still leaves the ADM with less function evaluations done in serial than the RK method of order p. This increase of order is easy to accomplish, as it is straight forward to obtain higher order ADM methods. These results support the statement, that the ADM can be a viable time stepping method.

It is mentioned, e.g. in [19], that the maximal time step width of the ADM could be increased with Padé's rational approximation. This might increase the viability of the ADM as an explicit time stepping scheme further while still allowing to exploit the additional degrees of parallelism in time.

Acknowledgements The work of Andreas Schmitt is supported by the 'Excellence Initiative' of the German Federal and State Governments and the Graduate School of Computational Engineering at Technische Universität Darmstadt.

This work was partially accomplished during a research stay at NCAR which provided the facilities.

References

1. Abbaoui, K., Cherruault, Y.: Convergence of Adomian's method applied to nonlinear equations. Math. Comput. Model. **20**(9), 69–73 (1994). https://doi.org/10.1016/0895-7177(94)00163-4
2. Abbasbandy, S., Darvishi, M.: A numerical solution of Burgers' equation by modified Adomian method. Appl. Math. Comput. **163**(3), 1265–1272 (2005). https://doi.org/10.1016/j.amc.2004.04.061
3. Adomian, G.: A new approach to nonlinear partial differential equations. J. Math. Anal. Appl. **102**(2), 420–434 (1984). https://doi.org/10.1016/0022-247X(84)90182-3
4. Adomian, G.: Nonlinear Stochastic Operator Equations. Academic, New York (1986)
5. Adomian, G.: Solution of physical problems by decomposition. Comput. Math. Appl. **27**(9), 145–154 (1994). https://doi.org/10.1016/0898-1221(94)90132-5
6. Adomian, G.: Solving Frontier Problems of Physics: The Decomposition Method. Springer Science+Business Media, Dordrecht (1994)
7. Adomian, G.: Explicit solutions of nonlinear partial differential equations. Appl. Math. Comput. **88**(2), 117–126 (1997). https://doi.org/10.1016/S0096-3003(96)00141-5
8. Akpan, I.: Adomian decomposition approach to the solution of the Burger's equation. Am. J. Comput. Math. **5**, 329–335 (2015). https://doi.org/10.4236/ajcm.2015.53030
9. Ascher, U.M., Ruuth, S.J., Spiteri, R.J.: Implicit-explicit Runge-Kutta methods for time-dependent partial differential equations. Appl. Numer. Math. **25**(2–3), 151–167 (1997). https://doi.org/10.1016/S0168-9274(97)00056-1
10. Barros, S.R., Peixoto, P.S.: Computational aspects of harmonic wavelet Galerkin methods and an application to a precipitation front propagation model. Comput. Math. Appl. **61**(4), 1217–1227 (2011)
11. Bateman, H.: Some recent researches on the motion of fluids. Mon. Weather Rev. **43**(4), 163–170 (1915). https://doi.org/10.1175/1520-0493(1915)43<163:SRROTM>2.0.CO;2
12. Burgers, J.: A mathematical model illustrating the theory of turbulence. Adv. Appl. Mech. **1**, 171–199 (1948). https://doi.org/10.1016/S0065-2156(08)70100-5
13. Butcher, J.C.: Implicit Runge-Kutta processes. Math. Comput. **18**(85), 50–64 (1964). https://doi.org/10.1090/S0025-5718-1964-0159424-9
14. Butcher, J.C.: Numerical Method for Ordinary Differential Equations, 2nd edn. Wiley, London (2008)
15. Cheng, M., Scott, K., Sun, Y., Wu, B.: Explicit solution of nonlinear electrochemical models by the decomposition method. Chem. Eng. Technol. **25**(12), 1155–1160 (2002). https://doi.org/10.1002/1521-4125(20021210)25:12<1155::AID-CEAT1155>3.0.CO;2-A
16. Cherruault, Y., Adomian, G.: Decomposition methods: a new proof of convergence. Math. Comput. Model. **18**(12), 103–106 (1993). https://doi.org/10.1016/0895-7177(93)90233-O
17. Cole, J.D.: On a quasi-linear parabolic equation occuring in aerodynamics. Q. Appl. Math. **9**, 225–236 (1951). https://doi.org/10.1090/qam/42889
18. Courant, R., Friedrichs, K., Lewy, H.: Über die partiellen Differenzengleichungen der mathematischen Physik. Math. Ann. **100**, 32–74 (1928). https://doi.org/10.1007/BF01448839
19. de Sousa Basto, M.J.F.: Adomian decomposition method, nonlinear equations and spectral solutions of Burgers equation. Ph.D. Thesis, Faculdade de Engenharia da Universidade do Porto (2006)
20. Durran, D.: Numerical Methods for Fluid Dynamics: With Applications to Geophysics. Texts in Applied Mathematics. Springer, New York (2010). https://doi.org/10.1007/978-1-4419-6412-0
21. El-Sayed, S.M., Kaya, D.: On the numerical solution of the system of two-dimensional Burgers' equations by the decomposition method. Appl. Math. Comput. **158**(1), 101–109 (2004). https://doi.org/10.1016/j.amc.2003.08.066

22. El-Tawil, M.A., Bahnasawi, A.A., Abdel-Naby, A.: Solving Riccati differential equation using Adomian's decomposition method. Appl. Math. Comput. **157**(2), 503–514 (2004). https://doi.org/10.1016/j.amc.2003.08.049
23. Gander, M.J.: 50 years of time parallel time integration. In: Multiple Shooting and Time Domain Decomposition. Springer, Berlin (2015). https://doi.org/10.1007/978-3-319-23321-5_3
24. Gottlieb, D., Orszag, S.: Numerical Analysis of Spectral Methods: Theory and Applications. CBMS-NSF Regional Conference Series in Applied Mathematics. Society for Industrial and Applied Mathematics, Philadelphia (1977)
25. Guellal, S., Grimalt, P., Cherruault, Y.: Numerical study of Lorenz's equation by the Adomian method. Comput. Math. Appl. **33**(3), 25–29 (1997). https://doi.org/10.1016/S0898-1221(96)00234-9
26. Hopf, E.: The partial differential equation $u_t + uu_x = \mu_{xx}$. Commun. Pure Appl. Math. **3**(3), 201–230 (1950). https://doi.org/10.1002/cpa.3160030302
27. Jiao, Y., Yamamoto, Y., Dang, C., Hao, Y.: An aftertreatment technique for improving the accuracy of Adomian's decomposition method. Comput. Math. Appl. **43**(6), 783–798 (2002). https://doi.org/10.1016/S0898-1221(01)00321-2
28. Kaya, D., Yokus, A.: A numerical comparison of partial solutions in the decomposition method for linear and nonlinear partial differential equations. Math. Comput. Simul. **60**(6), 507–512 (2002). https://doi.org/10.1016/S0378-4754(01)00438-4
29. Kaya, D., Yokus, A.: A decomposition method for finding solitary and periodic solutions for a coupled higher-dimensional Burgers equations. Appl. Math. Comput. **164**(3), 857–864 (2005). https://doi.org/10.1016/j.amc.2004.06.012
30. LeVeque, R.: Finite Difference Methods for Ordinary and Partial Differential Equations: Steady-State and Time-Dependent Problems. SIAM e-books. Society for Industrial and Applied Mathematics (SIAM), Philadelphia (2007)
31. Press, W.H., Flannery, B.P., Teukolsky, S.A., Vetterling, W.T., et al.: Numerical Recipes, vol. 3. Cambridge University Press, Cambridge (1989)
32. Shawagfeh, N., Kaya, D.: Comparing numerical methods for the solutions of systems of ordinary differential equations. Appl. Math. Lett. **17**(3), 323–328 (2004). https://doi.org/10.1016/S0893-9659(04)90070-5
33. Vadasz, P., Olek, S.: Convergence and accuracy of Adomian's decomposition method for the solution of Lorenz equations. Int. J. Heat Mass Trans. **43**(10), 1715–1734 (2000). https://doi.org/10.1016/S0017-9310(99)00260-4
34. Wesseling, P.: Principles of Computational Fluid Dynamics. Springer Series in Computational Mathematics. Springer, Berlin (2009)
35. Zhu, H., Shu, H., Ding, M.: Numerical solutions of two-dimensional Burgers' equations by discrete Adomian decomposition method. Comput. Math. Appl. **60**(3), 840–848 (2010). https://doi.org/10.1016/j.camwa.2010.05.031

Integrated Modeling and Validation for Phase Change with Natural Convection

Kai Schüller, Benjamin Berkels, and Julia Kowalski

Abstract Melting water-ice systems develop complex spatio-temporal interface dynamics and a non-trivial temperature field. In this contribution, we present computational aspects of a recently conducted validation study that aims at investigating the role of natural convection for cryo-interface dynamics of water-ice. We will present an established fixed grid model known as the enthalpy porosity method (Brent et al., Numer Heat Transf A 13(3):297–318, 1988; Kumar and Krishna, Energy Procedia 109:314–321, 2017). It is based on introducing a phase field and employs mixture theory. The resulting PDEs are solved using a finite volume discretization. The second part is devoted to experiments that have been conducted for model validation. The evolving water-ice interface is tracked based on optical images that show both the water and the ice phase. To segment the phases, we use a binary Mumford Shah method, which yields a piece-wise constant approximation of the imaging data. Its jump set is the reconstruction of the measured phase interface. Our combined simulation and segmentation effort finally enables us to compare the modeled and measured phase interfaces continuously. We conclude with a discussion of our findings.

Keywords Phase change · Finite volume method · OpenFOAM · Image segmentation

Nomenclature

A Kozeny-Carman relation
c_p heat capacity
$\bar{c}_p$ averaged heat capacity

K. Schüller (✉) · B. Berkels · J. Kowalski
AICES Graduate School, RWTH Aachen University, Aachen, Germany
e-mail: schueller@aices.rwth-aachen.de

M. Schäfer et al. (eds.), *Recent Advances in Computational Engineering*, Lecture Notes in Computational Science and Engineering 124, https://doi.org/10.1007/978-3-319-93891-2_8

c_1, c_2	gray scale values
C	mushy zone constant
f	phase mass fraction
$\mathbf{F}$	phase interaction force
$\mathbf{g}$	gravitational acceleration
h	enthalpy
h_m	latent heat of melting
k	thermal conductivity
p	pressure
$\mathbf{S}$	Boussinesq term
T	temperature
T_S	solidus temperature
T_L	liquidus temperature
T_m	melting temperature
T_{init}	initial PCM temperature
T_w	wall temperature
$\mathbf{u}$	velocity field
V	volume
γ	phase volume fraction
ϵ	small constant
η	dynamic viscosity
Θ	temperature deviation($\Theta = T - T_m$)
ρ	density
$\bar{\rho}$	partial density

1 Introduction

Phase change processes play an important role in a variety of present-day research fields and industrial applications. A material that undergoes phase change, a so-called phase change material (PCM), absorbs and releases heat at a constant temperature T_m or within a certain phase change temperature range, bounded by the liquidus temperature T_L and the solidus temperature T_S. PCMs are particularly relevant to thermal energy storage (TES) systems, because of their large storage density compared to non-latent TES systems (5–14 times more heat per unit volume than sensible storage materials [19]). A TES system is an attractive technology because it is the most appropriate method to correct the gap between demand and supply of energy [1]. This becomes very important in the context of renewable energy sources, because most of them depend on time-varying environmental parameters, such as the wind speed (for wind power plants) or the duration of solar irradiation (for solar power plants). TES systems are also used for cooling applications, e.g. to protect electrical devices. The cheapest PCM for cooling applications is water-ice. Its melting temperature is 0 °C. Beyond this industrial application, the process of water-ice melting can be found in a variety of scientific areas, e.g. glaciology or ice sheet modeling.

To simulate phase change heat transfer both the sensible and the latent heat release or storage must be considered, which translates into a moving boundary problem as the interface might propagate or retrieve. Such problems can be solved either with fixed- or deforming grid methods, or a combination of both [20]. Even though deforming grid methods are in general more accurate than fixed grid methods in terms of localizing the phase interface, fixed-grid methods are computationally much more efficient and allow to represent topological changes with ease. The major advantage of fixed grid methods is that the numerical treatment of the phase change can be achieved through simple modifications of existing numerical methods, which allows to model phase change for a variety of complex phase change systems with relative ease [22]. When the liquid phase of the PCM is convecting, the fluid flow can have a considerable impact on the heat transfer within the system. Therefore, it is necessary to both solve for the heat transfer and the fluid flow. A popular method that is used for such phase change processes is the so-called enthalpy porosity method [4, 13].

Unfortunately, there exists no analytical solution to verify phase change models with natural convection. However, one-dimensional phase change without natural convection can be verified by comparison to the analytical solution of the Stefan problem, which has been already done with great success for the enthalpy-porosity method [12]. To validate phase change with natural convection, experiments must be used. A very common benchmark is the melting of a PCM, which is driven by an isothermal vertical wall in a rectangular cavity. The majority of these experiments include PCMs with a melting temperature higher than 0 °C. Examples are gallium [8] and n-octadecane [9]. Similar experiments exist for water-ice [18].

In this contribution, we present a fixed grid model that uses the enthalpy porosity method to simulate phase change with natural convection. We implemented this established model into an OpenFOAM solver. In order to validate the model, we conducted our own experiments, which are similar to existing benchmark tests but with high spatio-temporal resolution. The data consist of optical images that show the motion of the phase interface over time. To extract the phase interface from the optical images, we use binary Mumford-Shah segmentation. This allows for a quantitative comparison between the model and the experimental results.

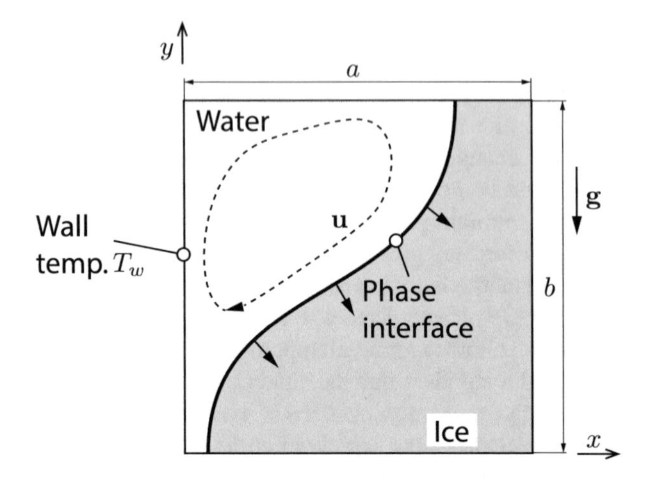

Fig. 1 Schematic of the physical situation. A two-dimensional cavity is filled with a PCM, which is present both in liquid, as well as its solid phase. Both phases are separated by the phase interface. The left boundary is held at constant temperature T_w

2 Model

2.1 Physical Situation

The physical situation is sketched in Fig. 1. A two-dimensional cavity of size $a \times b$ is filled with an initially solid phase change material (PCM) of temperature T_{init}. Due to an imposed temperature T_w at the left boundary, which is higher than the melting temperature of the PCM T_m, the PCM heats up locally and changes its phase from solid to liquid. Both phases are separated by a phase interface. The shape of the phase interface is mainly defined by natural convection. In the presence of gravitational acceleration **g**, the density variation in the liquid phase induces natural convection, which manifests as a clockwise rotational flow field **u** within the liquid phase. The presented approach is applicable to a variety of PCMs, e.g. metals or waxes. In this study, we will however focus on water-ice.

2.2 Model Equations

To formulate a fixed-grid mathematical model that describes the physical problem of phase change with natural convection, either volume-averaging or classical mixture theory can be utilized. Here, we will shortly sketch the latter approach based on mixture theory. Interested readers can find a comprehensive derivation of the mixture equations using volume-averaging in [15]. The basic idea of mixture theory is to introduce a scalar field, which stores the information of the PCM state.

Following [2], we use the phase volume fraction, which is defined as

$$\gamma_i = \frac{V_i}{\sum_i V_i} \tag{1}$$

in which V_i is the volume of phase i in a control volume. From Eq. (1), it can be seen that the value of the phase volume fraction is always between zero and unity. We further assume full saturation, i.e.

$$\sum_i \gamma_i = 1 \tag{2}$$

The partial density of phase i is then given by

$$\bar{\rho}_i = \gamma_i \rho_i \tag{3}$$

in which ρ_i is the density of phase i. The mass fraction of phase i is

$$f_i = \frac{\bar{\rho}_i}{\sum_i \bar{\rho}_i} \tag{4}$$

In this work, we are interested in a two-phase system, which is given by a solid and a liquid phase. Substituting Eqs. (2) and (3) into the mass fraction (4) yields an explicit relation for the liquid phase

$$f_L = \frac{\gamma_L \rho_L}{\gamma_L \rho_L + \gamma_S \rho_S} \tag{5}$$

In the special case of $\rho_L = \rho_S$, Eq. (5) reduces to $f_L = \gamma_L$ and analogously $f_S = \gamma_S$.

According to [2], the three mixture balance laws are obtained by summing the balance laws for the individual phases, i.e. conservation of mass, momentum and energy for the liquid and solid phase. After some simplifications and introducing a set of mixture variables and parameters, a system that accounts for incompressible mixture flow and phase change coupled to natural convection can be derived. It is given by

$$\nabla \cdot \mathbf{u} = 0 \tag{6}$$

$$\frac{\partial (\rho \mathbf{u})}{\partial t} + \nabla \cdot (\rho \mathbf{u} \otimes \mathbf{u}) = -\nabla p + \nabla \cdot (\eta \nabla \mathbf{u}) + \mathbf{F} + \mathbf{S}(T) \tag{7}$$

$$\frac{\partial (\rho h)}{\partial t} + \nabla \cdot (\rho \mathbf{u} h) = \nabla \cdot (k \nabla T) \tag{8}$$

in which

$$\rho = \gamma_S \rho_S + \gamma_L \rho_L \tag{9}$$

$$k = \gamma_S k_S + \gamma_L k_L \tag{10}$$

$$h = f_S h_S + f_L h_L \tag{11}$$

$$\eta = \eta_L \tag{12}$$

$$\mathbf{u} = \mathbf{u}_S = \mathbf{u}_L \tag{13}$$

are the mixture density ρ, mixture thermal conductivity k, mixture enthalpy h, mixture dynamic viscosity η and mixture velocity $\mathbf{u}$. Note that in the local presence of both phases, we assume them to move at the same velocity. This assumption is appropriate, because relative phase motion can be neglected.

Equation (7) is the conservation of momentum, which includes two additional terms, namely a temperature dependent Boussinesq approximation term $\mathbf{S}(T)$ and a phase interaction force term $\mathbf{F}$. The Boussinesq approximation term accounts for natural convection due to buoyancy and is defined as

$$\mathbf{S}(T) = \mathbf{g}\rho(T) \tag{14}$$

in which $\rho(T)$ is a polynomial fit to tabulated density data. It should be noted that the Boussinesq approximation is only valid if the density variation is small, which is a valid assumption for water.

The phase interaction force $\mathbf{F}$ accounts for momentum production due to phase interactions [2]. According to [21], the flow regime within cells that contain portions of both phases can be interpreted as a porous medium. Hence, the flow can be described by Darcy's law. This behavior can be accounted for by defining

$$\mathbf{F} = A\mathbf{u} \tag{15}$$

A is large in the liquid phase ($\gamma_L = 1$) and small in the solid phase ($\gamma_L = 0$). This allows for flow in the liquid phase, whereas it suppresses it in the solid phase. A commonly used continuous function with this properties is the Kozeny-Carman relation [21]

$$A = -C\frac{(1-\gamma_L)^2}{\gamma_L^3 + \epsilon} \tag{16}$$

Here, ϵ (typically $\epsilon = 10^{-6}$) is a stabilizing parameter that is used in order to prevent division by zero and C denotes the mushy zone constant. It should be noted that C has no direct physical significance and has to be calibrated with data. In non-isothermal phase change processes the PCM develops a mushy region rather than a sharp phase interface. In this case, adjusting the mushy zone constant can be exploited to model the resulting porosity near the mushy phase interface.

2.3 Source-Based Method for Phase Change

In order to solve the energy equation, which is given in enthalpy form, we need to introduce an equation that relates the enthalpy to the temperature. The enthalpies of the solid and liquid phases are given by

$$h_S = \int_{T_m}^{T} c_{p,S} \mathrm{d}T \tag{17}$$

$$h_L = \int_{T_m}^{T} c_{p,L} \mathrm{d}T + h_m \tag{18}$$

in which h_m is the latent heat of melting and $c_{p,S}$ as well as $c_{p,L}$ are the heat capacities of the solid and liquid phase, respectively.

Over a temperature range of 20 K, the percentage heat capacity change is in the order of 5 % for ice and 1 % for water. If we assume phase-wise constant heat capacities, which is a valid approximation as long as the temperature range within the PCM is small, Eqs. (17) and (18) simplify to

$$h_S = \bar{c}_{p,S}(T - T_m) \tag{19}$$

$$h_L = \bar{c}_{p,L}(T - T_m) + h_m \tag{20}$$

From Eq. (5), it can be seen that if the densities of the solid and liquid phases are equal, the mass fraction (4) has the same value as the volume fraction, i.e. $f_k = \gamma_k$. The difference between water and ice density is in the order of 10 %, i.e. the mass fraction roughly equals the volume fraction. Due to simplicity and because we do not expect the result to be qualitatively different, we will restrict ourselves to $\rho_S \approx \rho_L$ and substitute the mass fraction in the mixture enthalpy Eq. (11) by the volume fraction, which yields

$$h = \gamma_S h_S + \gamma_L h_L \tag{21}$$

Substituting the approximations for the solid (19) and liquid enthalpy (20) into the equation for the mixture enthalpy (21) yields

$$h = \bar{c}_p(T - T_m) + \gamma_L h_m \tag{22}$$

in which $\bar{c}_p = \gamma_L \bar{c}_{p,L} + \gamma_S \bar{c}_{p,S}$ is the mixture heat capacity. We can now substitute the mixture enthalpy (22) into the energy equation, which yields

$$\frac{\partial(\rho \bar{c}_p \Theta)}{\partial t} + \nabla \cdot \left(\rho \mathbf{u} \bar{c}_p \Theta\right) = \nabla \cdot (k \nabla \Theta) - h_m \left(\frac{\partial\left(\rho \gamma_L\right)}{\partial t} + \nabla \cdot (\rho \mathbf{u} \gamma_L) \right) \tag{23}$$

in which $\Theta = T - T_m$ denotes the deviation from the melting temperature.

The left-hand side and the first term of the right-hand side of Eq. (23) matches the standard transient convection-diffusion energy equation that describes sensible heat transfer. The remaining term accounts for the latent heat transfer due to phase change.

2.4 Solution Algorithm for the Energy Equation

Equation (23) contains two unknowns, namely the temperature Θ and the liquid volume fraction γ_L. These two fields, however, are intrinsically coupled. In order to solve Eq. (23), a relation between the temperature and the liquid volume fraction is required. In our work we follow [17] and use a piecewise linear function

$$\gamma_L = \begin{cases} 0, & T < T_S \\ \frac{T-T_S}{T_L-T_S}, & T_S \leq T \leq T_L \\ 1, & T > T_L \end{cases} \tag{24}$$

This approach assumes that the phase change occurs within a narrow temperature range $T_L - T_S$, rather than at a fixed temperature.

Following [21], we linearize Eq. (23) and introduce an iterative corrector approach

$$\frac{\partial(\rho\bar{c}_p\Theta^{k+1})}{\partial t} + \nabla\cdot\left(\rho\mathbf{u}\bar{c}_p\Theta^{k+1}\right) = \nabla\cdot\left(k\nabla\Theta^{k+1}\right) - h_m\left(\frac{\partial\left(\rho\gamma_L^k\right)}{\partial t} + \nabla\cdot\left(\rho\mathbf{u}\gamma_L^k\right)\right) \tag{25}$$

in which γ_L^k is the known volume fraction of the previous iteration k and Θ^{k+1} is the solution variable of the current iteration. The updated temperature Θ^{k+1} does not match the temperature determined through relation (24) based on the volume fraction of the previous iteration k. Therefore, an energy conserving updating of the volume fraction is used [7, 12]

$$\gamma_L^{k+1} = \max\left[\min\left[\gamma_L^k + \lambda\frac{\bar{c}_p}{h_m}\left(\Theta^{k+1} - \Theta_{\text{cons}}^{k+1}\right), 1\right], 0\right] \tag{26}$$

with

$$\Theta_{\text{cons}}^{k+1} = T_S + (T_L - T_S)\,\gamma_L^k - T_m \tag{27}$$

in which λ is a relaxation factor. According to [22], values between 0.5 and 0.7 provide efficient convergence for both one- and two-dimensional problems. The consistent temperature equation (27) directly follows from the volume fraction temperature relation (24). Equation (26) further assures that no over- and undershooting of the volume fraction occurs, i.e. the values will be always between zero and unity.

2.5 Summary of the Iterative Solution Procedure

Incorporating the stated equations, the following iterative solution procedure is applied to solve the energy equation (23):

1. Either set an initial liquid volume fraction γ_L^k if it is the first time step or use the volume fraction of the previous time step.
2. Solve the linearized energy equation (25) for Θ^{k+1}.
3. Calculate the temperature $\Theta_{\text{cons}}^{k+1}$, which is consistent to the volume fraction from the previous iteration using Eq. (27).
4. Update the volume fraction γ_L^{k+1} using Eq. (26). Go back to step 2 if the convergence threshold is not reached, i.e. if the error of the volume fraction is not smaller than a certain tolerance.

2.6 Implementation

For this work we used OpenFOAM, which is an object oriented open source C++ library to solve PDEs [10, 23]. We implemented the enthalpy porosity method by extending *buoyantBoussinesqPimpleFoam* (OpenFOAM 5.0), which is a transient solver for buoyant, turbulent flow of incompressible fluids that uses the PIMPLE algorithm for pressure velocity coupling.

2.7 Mushy Zone Constant Sensitivity

The sensitivity of the mushy zone constant with respect to the resulting phase interface has been studied for gallium [13] and for lauric acid [11]. Both studies conclude that the mushy zone constant significantly influences the shape of the resulting phase interface. So it should be chosen carefully in order to obtain reasonable results. To our knowledge, such sensitivity studies have not been conducted for water-ice PCMs. Since we want to validate our phase change simulations against water-ice PCMs, we also studied the results for different mushy zone constants. Figure 2 shows the phase interfaces at 600 and 900 s. The simulations have been conducted on a quadratic uniform mesh of 102,400 quadrilateral cells using adiabatic boundaries, except for the left boundary at which a Dirichlet condition of 30.5 °C is applied. Furthermore, we use no-slip conditions at all boundaries and temperature dependent thermophysical material properties. The initial temperature is $T_{\text{init}} = -20\,°\text{C}$. It can be seen, that the phase interface oscillates for a mushy zone constant of $C = 10^6$. Increasing the mushy zone constant yields a smoother phase interface. Furthermore, the plot shows that the phase interface converges if the mushy zone constant is increased. Based on our findings, we chose a value of $C = 10^{10}$ for all following simulations.

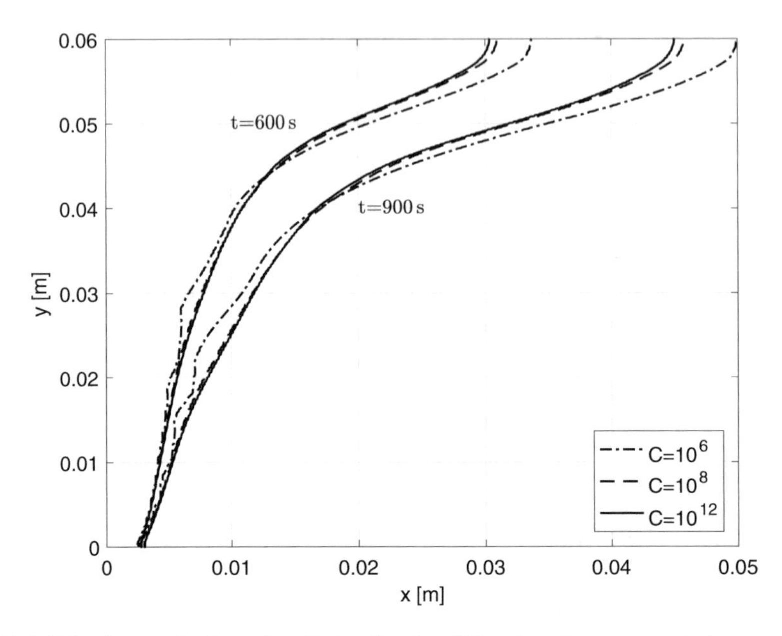

Fig. 2 Melting from an isothermal vertical wall using different mushy zone constants

3 Validation Experiment

3.1 Apparatus and Instrumentation

The experimental setup is shown in Fig. 3. It consists of a PCM container, which is made out of plexiglas and one optical as well as one infrared camera. In order to observe the melting process, the PCM container has two circular windows of different materials, namely plexiglas for the optical camera and germanium for the infrared camera. In this work, we will focus on the results of the optical camera. The PCM container is equipped with two heater blocks of size $30 \times 60 \times 20\,\text{mm}^3$ that are in contact with the PCM. Each heater block contains two heating cartridges, which can be controlled independently. The heater blocks contain temperature sensors for temperature control. A feedback loop sustains a predefined temperature by means of a straight-forward feedback control.

For this work, we will study the case of an isothermal vertical wall. So we do not use the bottom heater. The dimensions of the inner PCM container are $30 \times 107 \times 114\,\text{mm}^3$.

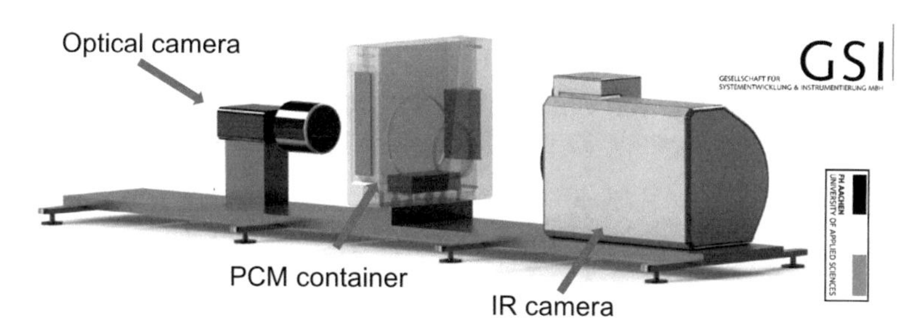

Fig. 3 Experiment assembly

3.2 Experimental Procedure

Before each experiment, the PCM container was removed from the experiment assembly, which was located in an approximately 22 °C warm laboratory. The container was filled with liquid water and put into a freezer in order to transform the water into ice. The water level was always above the side heater before freezing. To obtain better ice qualities, a low power heat source has been placed close to the water surface, so that the solidification proceeds from the bottom to the top of the container. Otherwise high stresses could have damaged the PCM container. After a certain amount of time, the ice temperature was approximately −20 °C. Then, the container has been removed from the freezer and reassembled into the experimental setup. Then, the heater configuration including the target temperature was set. The logging of data started together with the activation of the vertical heater block. A switch-on temperature of 29 °C and a switch-off temperature of 30 °C in the feedback control led to temperatures of the heater block oscillating between 28 °C and 33 °C due to thermal inertia of the heater blocks.

4 Image Segmentation

To extract the water-ice interface from the optical images, we use the concept of image segmentation. Since there are just two different segments (water and ice), we are facing a two-phase image segmentation problem.

Let $\Omega \subset \mathbb{R}^2$ denote the image plane. Given an image $g : \Omega \to \mathbb{R}$, we are searching for a piecewise constant segmentation, i.e. two gray values c_1, c_2 and a region $\mathcal{O} \subset \Omega$ that minimizes

$$E[\mathcal{O}, c_1, c_2] = \int_{\mathcal{O}} (g - c_1)^2 \, dx + \int_{\Omega \setminus \mathcal{O}} (g - c_2)^2 \, dx + \nu \mathrm{Per}(\mathcal{O}).$$

Here, $\mathrm{Per}(\mathcal{O})$ is the perimeter of $\mathcal{O}$, i.e. the length of the phase interface. For a fixed $\mathcal{O}$, the optimal gray values c_1 and c_2 are just the average values of g inside $\mathcal{O}$ and $\Omega \setminus \mathcal{O}$ respectively. The minimization with respect to $\mathcal{O}$ is difficult. Denoting $f_i := (g - c_i)^2$, we consider the so-called binary Mumford-Shah functional [14]

$$E_{\mathrm{MS}}[\mathcal{O}] = \int_{\mathcal{O}} f_1 \,\mathrm{d}x + \int_{\Omega \setminus \mathcal{O}} f_2 \,\mathrm{d}x + \nu \mathrm{Per}(\mathcal{O}).$$

Minimizing E_{MS} is a nonconvex optimization problem, since the set of subsets of Ω is not convex. Fortunately, a strongly convex reformulation of this problem is available. The main idea is to replace $\mathcal{O}$ by a function $w : \Omega \to \mathbb{R}$. This leads to the functional

$$E_{\mathrm{UC}}[w] = \int_{\Omega} w^2 f_1 \,\mathrm{d}x + \int_{\Omega} (1 - w)^2 f_2 \,\mathrm{d}x + \nu \mathrm{TV}[w],$$

where $\mathrm{TV}[w]$ denotes the Total Variation of w. Denoting by $\chi_{\mathcal{O}}$ the characteristic function of $\mathcal{O}$, it is easy to show that $E_{\mathrm{MS}}[\mathcal{O}] = E_{\mathrm{UC}}[\chi_{\mathcal{O}}]$. In this sense, minimizing E_{UC} over $BV(\Omega)$, the set functions with finite Total Variation, is a relaxation of the problem to minimize E_{MS} over the subsets of Ω. The former is a strongly convex problem and as such its unique minimizer can be computed efficiently. Moreover, this minimizer encodes a minimizer of the original non-convex problem. One can show [3, 5] that

$$w^* = \operatorname*{argmin}_{w \in BV(\Omega)} E_{\mathrm{UC}}[w] \quad \Rightarrow \quad \{w^* > 0.5\} \in \operatorname*{argmin}_{\mathcal{O} \subset \Omega} E_{\mathrm{MS}}[\mathcal{O}]$$

where $\{w > 0.5\}$ is the 0.5-superlevel set of w, i.e. $\{x : w(x) > 0.5\}$. That means the optimization with respect to $\mathcal{O}$ can be solved by minimizing E_{UC} and thresholding the minimizer. The numerical optimization uses a dual formulation. Recall that the Total Variation is defined as

$$\mathrm{TV}[w] = \sup_{q \in K} \int_{\Omega} w \nabla \cdot q \,\mathrm{d}x$$

where $K = \left\{ q \in C_c^\infty(\Omega, \mathbb{R}^d) : |q(x)| \leq 1 \text{ for all } x \in \Omega \right\}$. Thus, E_{UC} can be minimized by solving a saddle point problem (minimizing in the primal variable w, maximizing in dual variable q). Efficient and simple first order algorithms for this are well known [6].

5 Results and Discussion

5.1 Image Segmentation Results

Figures 4 and 5 show the images taken from the experiment at 600 and 900 s, respectively. A small portion of the two heater blocks is visible on the left and at the bottom. The heater blocks have been used as a reference to scale the image pixels to the size of the experiment. It can be seen that the water is dark compared to the ice, which enabled us to apply our segmentation approach. To compensate for the somewhat non-uniform illumination inherent to our experimental setup, we estimated the background illumination of the scene by applying the morphological opening and closing operator to the first frame of the video and subtracted this illumination estimate from each video frame before applying the segmentation. Empirically, we found that initial gray values (c_1 and c_2) of 0.3 and 0.5 work best for the images, which were taken from the experiment. The result of the image segmentation is plotted on top of the images. Even though the phase interfaces have been detected very good, there are some small artifacts due to similar gray values, e.g. at the circumference of the window in Fig. 4. To better compare the experiment and our numerical results, we arbitrarily chose nine data points (plotted as circles), which are equidistant in y-direction.

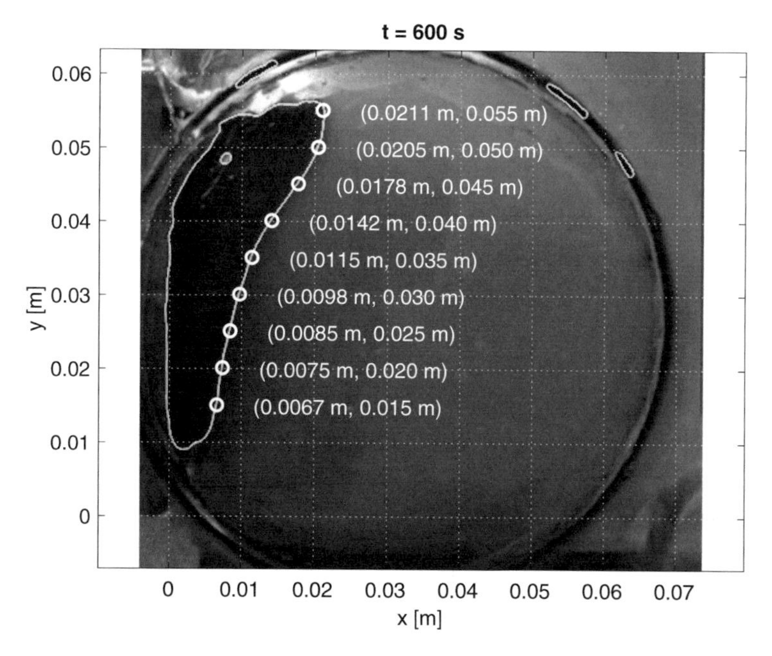

Fig. 4 Image taken from the experiment after 600 s and segmentation result, as well as the position of nine data points (circles)

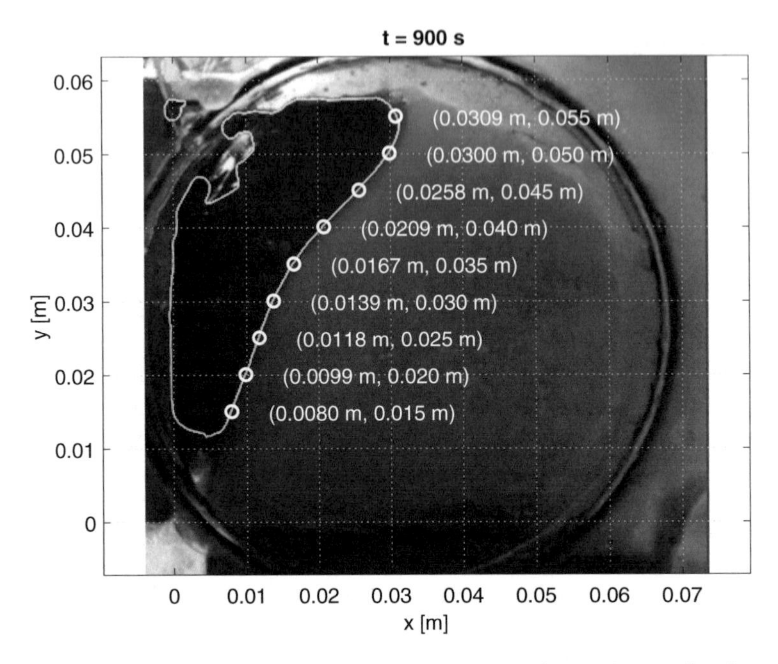

Fig. 5 Image taken from the experiment after 900 s and segmentation result, as well as the position of nine data points (circles)

The effect of natural convection due to buoyancy is clearly visible in the images. The heating at the left wall causes a decrease in density of the nearby water, which on the other hand induces a flow field in upward direction. Near the phase interface, heat is absorbed by the ice so that the density increases relative to the average temperature within the water phase. As the water flows in downward direction along the phase interface, it constantly cools down. As a consequence the temperature and hence the melting rate is higher near the top compared to the bottom. The whole process results in a circular flow field in clockwise direction.

5.2 Comparison with Experiment

In order to compare the experimental results to our model, we used a quadratic uniform mesh of 102,400 quadrilateral cells. Using a computational domain with the same size of the inner PCM container domain would be computationally inefficient, since most of the space is occupied by ice. Instead, we use a smaller computational domain of $0.06 \times 0.06\,\mathrm{m}^2$, which is large enough to include the entire water phase throughout the simulation. In order to use temperature dependent material properties for water-ice, we used approximations that fit tabular data from the literature, e.g.

from [16]. To give an example, the density of the water has been approximated using

$$\rho_L = \sum_{i=0}^{3} R_i \left(T - T_{ref}\right)^i + R_4 \left(T - T_{ref}\right)^{2.5} \tag{28}$$

in which $T_{ref} = 273.15\,\text{K}$, $R_0 = 999.79684\,\text{kg/m}^3$, $R_1 = 0.068317355\,\text{kg/m}^3/\text{K}$, $R_2 = -0.010740248\,\text{kg/m}^3/\text{K}^2$, $R_3 = -2.3030988 \times 10^{-5}\,\text{kg/m}^3/\text{K}^3$ and $R_4 = 0.00082140905\,\text{kg/m}^3/\text{K}^{2.5}$.

Except for the left wall at which a Dirichlet condition of 30.5 °C is applied, all boundaries are adiabatic. We further assigned no-slip conditions on all walls. The initial temperature was set to $T_{\text{init}} = -20\,°\text{C}$. The solidus and liquidus temperatures were set to $T_S = -0.05\,°\text{C}$ and $T_L = 0\,°\text{C}$, respectively. The mushy zone constant was set to $C = 10^{10}$ based on our findings in the sensitivity analysis. For the iterative solution of the energy equation, we used a tolerance of 10^{-8} for the liquid volume fraction.

Figure 6 shows the comparison of the phase interface positions of the experiment and the simulation at 600 and 900 s. The phase interface obtained by the simulation qualitatively fits the experimental results, even though there is a significant offset between both results. It can be seen that the maximum melting rate is located at the top for both the experiment and the simulation results. The maximum error is at the top ($y = 0.06\,\text{m}$). It is smaller at 600 s, at which the phase interface is captured really well, compared to the results at 900 s. At a height of approximately 0.05 m,

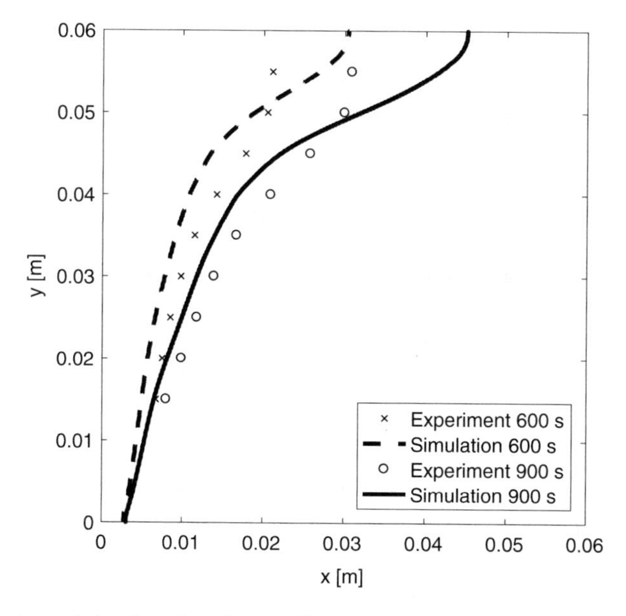

Fig. 6 Comparison of the phase interface positions of the experiment and the simulation at 600 and 900 s

the phase interface of the simulation and the experiment intersect and below 0.05 m the phase interface position of the simulation migrates slower than it was observed in the experiment.

Besides the fact that we simulated the process using a two-dimensional domain instead of a more realistic three-dimensional domain, there are some additional error sources and uncertainties, which could explain the discrepancy between the experiment and the numerical results. The initial temperature of the ice is inferred from the preparation procedure. However, we tested the range of possible initial temperatures between −25 °C and 0 °C and discovered only a small sensitivity. A larger error could result from too idealized boundary conditions. We assumed adiabatic walls, except for the boundary at which the heater is located. In the experiment, the water-ice PCM was in contact to plexiglas walls, which will introduce a heat sink, because the experiment has been conducted in a laboratory with an ambient temperature of 22 °C. The heater has been modeled by using a Dirichlet boundary condition. However, the heater temperature oscillates between 28 °C and 33 °C due to the control loop.

6 Conclusions and Outlook

In this contribution we describe an established fixed grid model to simulate phase change processes with natural convection. The model is based on the enthalpy porosity method, a phase field method, which can be derived from classical mixture theory. We use an iterative corrector approach to solve the resulting nonlinear energy equation. The final system of PDEs that describes the incompressible mixture flow with phase change has been solved using OpenFOAM. The method uses a parameter referred to as the mushy zone constant. A sensitivity study suggests that the mushy zone constant should be high in order to capture the physical regime of water-ice.

In order to validate the model, we conducted experiments in which water-ice was melted from an isothermal vertical wall. We tracked the water-ice interface using optical images, which resulted in experimental data of the phase interface at high spatio-temporal resolution. In order to utilize this data, we used the Mumford Shah method to segment the phases in the imaging data and to quantify the phase interface position. Our results demonstrate the proficiency of this approach for water-ice segmentation in images. It allows for comparison between the simulation and the experiment.

We observed good qualitative agreement regarding the shape throughout the whole evolution of the phase interface. Measured from the left boundary, the maximum distance of the phase interface is near the top, which directly follows from the buoyancy-induced flow field in the liquid phase. Although the results look qualitatively similar, there is, however, a an error between the simulation and experiment in terms of the phase interface position.

This inconsistency is still under investigation. Possible explanations include too idealized boundary conditions in the simulation and a bad insulation regarding the experiment. These must be investigated in the future, either by extending the simulation or by conducting tailored experiments using a redesigned experimental setup with less uncertainties than introduced by the present setup. Following various previous work, we assumed that the density difference between water and ice does not play a major role. The validity of this assumption also needs to be assessed more carefully in future work.

In general, both our capability to simulate complex multi-physics problems, as well as our capability to acquire data grew extensively in recent years. Optimal combination of both that result in high quality model validation strategies at high spatio-temporal resolution are, however, rare. Standard practice is often rather to compare sophisticated models to a sparse data set, or to analyze large data sets with very idealized models. Exceptions exist for certain processes, e.g. as relevant for meteorology, but cannot be easily extended to arbitrary process models. On our way to explore sophisticated model validation strategies at high spatio-temporal resolution, we proposed to set up a tailored laboratory experiment and designed data processing to match ideally with our major simulation goal. Inconsistencies between the simulation and experiment are accessible, which would be hard to acknowledge if validation had been done with sparse data only. Our next steps will be twofold, namely specifically investigating the inconsistencies between model and experiments in the concrete conducted validation study, and more generally continue to work on flexible, integrated validation strategies for coupled multi-physics systems.

Acknowledgements The project was funded in part by the Excellence Initiative of the German Federal and State Governments. It is supported by the Federal Ministry for Economic Affairs and Energy, Germany, on the basis of a decision by the German Bundestag (FKZ: 50 NA 1502). It is part of the Enceladus Explorer initiative of the DLR Space Administration.

References

1. Akeiber, H.J., Wahid, M.A., Hussen, H.M., Mohammad, A.T.H.: Review of development survey of phase change material models in building applications. Sci. World J. **2014**, 391690 (2014)
2. Bennon, W.D., Incropera, F.P.: A continuum model for momentum, heat and species transport in binary solid-liquid phase change systems—I. Model formulation. Int. J. Heat Mass Transf. **30**(10), 2161–2170 (1987)
3. Berkels, B.: An unconstrained multiphase thresholding approach for image segmentation. In: Proceedings of the Second International Conference on Scale Space Methods and Variational Methods in Computer Vision (SSVM 2009). Lecture Notes in Computer Science, vol. 5567, pp. 26–37. Springer, Berlin, (2009)
4. Brent, A.D., Voller, V.R., Reid, K.T.J.: Enthalpy-porosity technique for modeling convection-diffusion phase change: application to the melting of a pure metal. Numer. Heat Transf. A **13**(3), 297–318 (1988)
5. Chambolle, A., Darbon, J.: On total variation minimization and surface evolution using parametric maximum flows. Int. J. Comput. Vis. **84**(3), 288–307 (2009)

6. Chambolle, A., Pock, T.: A first-order primal-dual algorithm for convex problems with applications to imaging. J. Math. Imaging Vis. **40**(1), 120–145 (2011)
7. Faden, M., König-Haagen, A., Höhlein, S., Brüggemann, D.: An implicit algorithm for melting and settling of phase change material inside macrocapsules. Int. J. Heat Mass Transf. **117**, 757–767 (2018)
8. Gau, C., Viskanta, R.: Melting and solidification of a pure metal on a vertical wall. J. Heat Transf. **108**(1), 174–181 (1986)
9. Ho, C.-J., Viskanta, R.: Heat transfer during melting from an isothermal vertical wall. J. Heat Transf. **106**(1), 12–19 (1984)
10. Jasak, H., Jemcov, A., Tukovic, Z., et al.: OpenFOAM: a C++ library for complex physics simulations. In: International Workshop on Coupled Methods in Numerical Dynamics, vol. 1000, pp. 1–20. IUC Dubrovnik, Croatia (2007)
11. Kheirabadi, A.C., Groulx, D.: The effect of the mushy-zone constant on simulated phase change heat transfer. In: ICHMT Digital Library Online. Begel House Inc., Danbury (2015)
12. König-Haagen, A., Franquet, E., Pernot, E., Brüggemann, D.: A comprehensive benchmark of fixed-grid methods for the modeling of melting. Int. J. Therm. Sci. **118**, 69–103 (2017)
13. Kumar, M., Krishna, D.J.: Influence of mushy zone constant on thermohydraulics of a PCM. Energy Procedia **109**, 314–321 (2017)
14. Mumford, D., Shah, J.: Optimal approximations by piecewise smooth functions and associated variational problems. Commun. Pure Appl. Math. **42**(5), 577–685 (1989)
15. Ni, J., Beckermann, C.: A volume-averaged two-phase model for transport phenomena during solidification. Metall. Mater. Trans. B **22**(3), 349–361 (1991)
16. Popiel, C.O., Wojtkowiak, J.: Simple formulas for thermophysical properties of liquid water for heat transfer calculations (from 0 °C to 150 °C). Heat Transf. Eng. **19**(3), 87–101 (1998)
17. Rösler, F.: Modellierung und Simulation der Phasenwechselvorgänge in makroverkapselten latenten thermischen Speichern, vol. 24. Logos Verlag Berlin GmbH, Berlin (2014)
18. Schütz, W., Beer, H.: Melting of ice in pure and saline water inside a square cavity. Chem. Eng. Process. Process Intensif. **31**(5), 311–319 (1992)
19. Sharma, A., Tyagi, V.V., Chen, C.R., Buddhi, D.: Review on thermal energy storage with phase change materials and applications. Renew. Sust. Energ. Rev. **13**(2), 318–345 (2009)
20. Voller, V.R.: An overview of numerical methods for solving phase change problems. Adv. Numer. Heat Transf. **1**(9), 341–380 (1997)
21. Voller, V.R., Prakash, C.: A fixed grid numerical modelling methodology for convection-diffusion mushy region phase-change problems. Int. J. Heat Mass Transf. **30**(8), 1709–1719 (1987)
22. Voller, V.R., Swaminathan, C.R., Thomas, B.G.: Fixed grid techniques for phase change problems: a review. Int. J. Numer. Methods Eng. **30**(4), 875–898 (1990)
23. Weller, H.G., Tabor, G., Jasak, H., Fureby, C.: A tensorial approach to computational continuum mechanics using object-oriented techniques. Comput. Phys. **12**(6), 620–631 (1998)

A Weighted Reduced Basis Method for Parabolic PDEs with Random Data

Christopher Spannring, Sebastian Ullmann, and Jens Lang

Abstract This work considers a weighted POD-greedy method to estimate statistical outputs parabolic PDE problems with parametrized random data. The key idea of weighted reduced basis methods is to weight the parameter-dependent error estimate according to a probability measure in the set-up of the reduced space. The error of stochastic finite element solutions is usually measured in a root mean square sense regarding their dependence on the stochastic input parameters. An orthogonal projection of a snapshot set onto a corresponding POD basis defines an optimum reduced approximation in terms of a Monte Carlo discretization of the root mean square error. The errors of a weighted POD-greedy Galerkin solution are compared against an orthogonal projection of the underlying snapshots onto a POD basis for a numerical example involving thermal conduction. In particular, it is assessed whether a weighted POD-greedy solutions is able to come significantly closer to the optimum than a non-weighted equivalent. Additionally, the performance of a weighted POD-greedy Galerkin solution is considered with respect to the mean absolute error of an adjoint-corrected functional of the reduced solution.

Keywords Weighted reduced basis method · Uncertainty quantification · Model order reduction · Proper orthogonal decomposition · Weighted POD-greedy

C. Spannring (✉) · S. Ullmann · J. Lang
Graduate School of Computational Engineering, Technische Universität Darmstadt, Darmstadt, Germany

Department of Mathematics, Technische Universität Darmstadt, Darmstadt, Germany
e-mail: spannring@gsc.tu-darmstadt.de; http://www.graduate-school-ce.de/index.php?id=688

M. Schäfer et al. (eds.), *Recent Advances in Computational Engineering*, Lecture Notes in Computational Science and Engineering 124,
https://doi.org/10.1007/978-3-319-93891-2_9

1 Introduction

Because the computational complexity of numerical simulations is growing, model order reduction becomes an essential task. In the last decades the reduced basis method (RBM) was extensively developed, e.g. recent overviews can be found in [8, 10, 18]. Furthermore, RBM was applied to stochastically influenced parametrized partial differential equations (PPDEs). A review can be found in [4]. In [3] the idea of a weighted RBM for elliptic PDEs was introduced. It allows to build up more efficient reduced spaces regarding an approximation of statistical quantities. In the following the focus is on the approximation of the expectation. The question arises, how much faster does the expected error converge for a weighted approach and how close is the reduced solution to an optimal reduced solution. Using a proper orthogonal decomposition (POD) [12], an optimal reduced space concerning the mean square error can be constructed. Furthermore, by a standard primal-dual approach [17] also the expected error of a linear output functional is considered.

This work shows different ways to construct reduced order models (ROMs) based on the RBM and the POD. Both methods have their justification. On the one hand the RBM outperforms the POD in computational time. On the other hand the POD accomplishes that the mean square error becomes minimal. The ROM are assessed regarding the approximation of the expected value.

The work is organized as follows. In Sect. 2 the model problem is formulated, and the notation for the high dimensional discretization is introduced. In Sect. 3 the reduced model and its assumptions are described. Non-weighted and weighted error estimators for the reduced model are stated in Sect. 4. Further, the reduced space construction for the non-weighted and the weighted approach is described in Sect. 5. Section 6 describes a ROM, obtained by a Galerkin projection onto a POD. Numerical results for an instationary heat equation are presented in Sect. 7.

2 Parametrized Linear Parabolic Model

In the following the weak formulation of a parametrized, linear parabolic PDE is considered, where a p-dimensional parameter vector is random. Let $(\Theta, \mathcal{F}, \mathbb{P})$ be a complete probability space where the sample space Θ contains all possible outcomes $\theta \in \Theta$. The sigma algebra $\mathcal{F}$ is given by a subset of all possible subsets of Θ, i.e. $\mathcal{F} \subseteq 2^{\Theta}$, and the probability measure $\mathbb{P}\colon \mathcal{F} \to [0, 1]$ maps an event to its probability. Let $\xi : \Theta \to \Gamma$ denote a random parameter vector whose image lies in a given parameter domain $\Gamma \subset \mathbb{R}^p$, which is determined by the support of the random variables. We assume that ξ has a joint probability density function (pdf) $\rho\colon \Gamma \to \mathbb{R}^+$ such that $\int_\Theta \mathrm{d}\mathbb{P}(\theta) = \int_\Gamma \rho(\xi)\mathrm{d}\xi = 1$.

The interest lies not only in the solution of the parabolic PDE problem itself but rather in some parameter-dependent output $s : \Gamma \to \mathbb{R}$. It is computed by an linear (output) functional $l\colon X \to \mathbb{R}$ that maps the solution, lying in a Sobolev space X, to a scalar output. The continuous parametrized problem reads as follows: For given $\xi \in \Gamma$, compute

$$s(\xi) = l(u(T; \xi)),$$

where for any $t \in [0, T]$ the solution $u(t; \xi) \in X$ fulfills

$$\begin{aligned} \langle \partial_t u(t; \xi), v \rangle + a(u(t; \xi), v) &= b(v; \xi), \quad && \forall v \in X, \\ (u(0; \xi), v)_{L^2(\Omega)} &= 0, && \forall v \in X. \end{aligned}$$

Here, $\langle \cdot, \cdot \rangle$ denotes a duality pairing between X' and X. The time derivative of the solution needs to lie in the dual space, i.e. $\partial_t u(t; \xi) \in X'$ for all parameters. In order to guarantee existence and uniqueness of the problem, the bilinear form is uniformly coercive and uniformly bounded and the functionals are uniformly bounded, i.e.

$$\overline{\alpha} \leq \alpha(\xi) = \inf_{0 \neq v \in X} \frac{a(v, v; \xi)}{\|v\|_X^2}, \qquad \sup_{0 \neq u, v \in X} \frac{a(u, v; \xi)}{\|u\|_X \|v\|_X} = \gamma(\xi) \leq \overline{\gamma}, \text{ and} \tag{1}$$

$$\sup_{0 \neq v \in X} \frac{|b(v; \xi)|}{\|v\|_X} = \gamma_b(\xi) \leq \overline{\gamma}_b, \qquad \sup_{0 \neq v \in X} \frac{|l(v)|}{\|v\|_X} = \gamma_l \leq \overline{\gamma}_l, \tag{2}$$

with $0 < \overline{\alpha}$ and $\overline{\gamma}, \overline{\gamma}_b, \overline{\gamma}_l < \infty$, for all $\xi \in \Gamma$. Furthermore, the bilinear form is assumed to be symmetric, i.e. $a(u, v; \cdot) = a(v, u; \cdot)$, $\forall u, v \in X$. It defines a parameter-dependent and parameter-independent energy norm, such that

$$\|v\|_\xi = \sqrt{(v, v)_\xi} = \sqrt{a(v, v; \xi)}, \qquad \forall v \in X, \forall \xi \in \Gamma, \tag{3}$$

$$\|v\|_{\xi_{\text{ref}}} = \sqrt{(v, v)_{\xi_{\text{ref}}}} = \sqrt{a(v, v; \xi_{\text{ref}})}, \quad \forall v \in X. \tag{4}$$

The parameter-independent norm is determined by a fixed reference parameter $\xi_{\text{ref}} \in \Gamma$.

2.1 *Fully Discretized Problem*

The problem is discretized with an implicit Euler method in time and with a linear finite element method in space. The time interval $[0, T]$ is split into $K \in \mathbb{N}$ equidistant time intervals with time step size $\Delta t := \frac{T}{K}$. The solution is approximated at time points $\{t^k = k\Delta t : k = 0, \ldots, K\}$, such that $u(x, t^k; \xi) \approx u^k(x; \xi)$. For the space discretization, h denotes the spatial step size and the high dimensional space $X_h := \text{span}\{\phi_1, \ldots, \phi_{\mathcal{N}}\} \subset X$, with $\dim(X_h) = \mathcal{N}$, contains piecewise linear basis functions. Further, $X_h^{K+1} := X_h \times \cdots \times X_h$ denotes the $(K + 1)$th power of the discretized space. The fully discretized primal problem reads as follows: For given

$\xi \in \Gamma$, compute

$$s_h(\xi) = l(u_h^K(\xi)), \tag{5}$$

where the detailed solutions $\{u_h^k(\xi)\}_{k=0}^K \in X_h^{K+1}$ fulfill

$$(u_h^k(\xi), v)_{L^2(\Omega)} + \Delta t\, a(u_h^k(\xi), v; \xi) \\ = (u_h^{k-1}(\xi), v)_{L^2(\Omega)} + \Delta t\, b(v; \xi), \quad \forall v \in X_h,\ k = 1, \ldots, K, \tag{6}$$

$$(u_h^k(\xi), v)_{L^2(\Omega)} = 0, \qquad \forall v \in X_h,\ k = 0. \tag{7}$$

A finite element method entails an $\mathcal{N}$-dimensional system of linear algebraic equations and hence the discretized problem is computationally expensive to solve. The solution coefficients, coming out of the algebraic equations, uniquely represent the detailed solution, such that

$$u_h^k(\xi) = \sum_{i=1}^{\mathcal{N}} u_{h,i}^k(\xi)\phi_i, \quad k = 0, \ldots, K.$$

Equations (5)–(7) are referred to as the detailed model and $u_h^k(\cdot)$ as the detailed solution. It can be seen as a reference solution. In the following the error is measured between the solution of the detailed model and the reduced model.

In order to achieve higher accuracy for the output computation, a dual (or adjoint) problem is used [6, Chapter 2.1]. This is a standard approach in the context of error analysis for functionals [15]. The dual problem of (5)–(7) reads as follows: For given $\xi \in \Gamma$, find the dual solutions $\{\psi_h^k(\xi)\}_{k=0}^K \in X_h^{K+1}$, s.t.

$$(v, \psi_h^k(\xi))_{L^2(\Omega)} + \Delta t\, a(v, \psi_h^k(\xi); \xi) \\ = (v, \psi_h^{k+1}(\xi))_{L^2(\Omega)}, \quad \forall v \in X_h,\ k = 0, \ldots, K-1, \tag{8}$$

$$(v, \psi_h^k(\xi))_{L^2(\Omega)} = l(v), \qquad \forall v \in X_h,\ k = K. \tag{9}$$

Due to the parameter-independent functional $l(\cdot)$, the solution of the final condition in (9) is parameter-independent, i.e. $\psi_h^K(\xi) = \psi_h^K$. Note, the dual problem evolves backward in time, hence the solutions are computed for decreasing time index k. As for the primal problem, the dual solution has a unique representation

$$\psi_h^k(\xi) = \sum_{i=1}^{\mathcal{N}} \psi_{h,i}^k(\xi)\phi_i, \quad k = 0, \ldots, K.$$

3 Reduced Basis Method

In order to decrease the computation time, a ROM is sought. This can be achieved using the RBM [10, 18]. It is a Galerkin projection onto a reduced basis space $X_N := \text{span}\{\varphi_1, \dots, \varphi_N\} \subset X_h$. The basis functions are orthogonal w.r.t. (4) and the dimension is denoted by $\dim(X_N) = N$. Additionally, a reduced space for the dual problem $\widetilde{X}_{\tilde{N}} := \text{span}\{\zeta_1, \dots, \zeta_{\tilde{N}}\} \subset X_h$ is introduced. The basis functions are orthogonal w.r.t. (4) and the dimension is denoted by $\dim(\widetilde{X}_{\tilde{N}}) = \tilde{N}$. The $(K+1)$th power of the reduced spaces are defined by $X_N^{K+1} := X_N \times \cdots \times X_N$ and $\widetilde{X}_{\tilde{N}}^{K+1} := \widetilde{X}_{\tilde{N}} \times \cdots \times \widetilde{X}_{\tilde{N}}$ respectively. Then, the reduced primal problem reads as follows: For given $\xi \in \Gamma$, find the reduced solutions $\{u_N^k(\xi)\}_{k=0}^K \in X_N^{K+1}$, s.t.

$$(u_N^k(\xi), v)_{L^2(\Omega)} + \Delta t\, a(u_N^k(\xi), v; \xi) = (u_N^{k-1}(\xi), v)_{L^2(\Omega)} + \Delta t\, b(v; \xi), \quad \forall v \in X_N,\ k = 1, \dots, K, \tag{10}$$

$$(u_N^k(\xi), v)_{L^2(\Omega)} = 0, \qquad \forall v \in X_N,\ k = 0. \tag{11}$$

The reduced dual problem reads as follows: For given $\xi \in \Gamma$, find the reduced solutions $\{\psi_{\tilde{N}}^k(\xi)\}_{k=0}^K \in \widetilde{X}_{\tilde{N}}^{K+1}$, s.t.

$$(v, \psi_{\tilde{N}}^k(\xi))_{L^2(\Omega)} + \Delta t\, a(v, \psi_{\tilde{N}}^k(\xi); \xi) = (v, \psi_{\tilde{N}}^{k+1}(\xi))_{L^2(\Omega)}, \quad \forall v \in \widetilde{X}_{\tilde{N}},\ k = 0, \dots, K-1, \tag{12}$$

$$(v, \psi_{\tilde{N}}^k)_{L^2(\Omega)} = l(v), \qquad \forall v \in \widetilde{X}_{\tilde{N}},\ k = K. \tag{13}$$

Note, the reduced spaces X_N and $\widetilde{X}_{\tilde{N}}$ are spanned by a different basis and in general they can have different dimensions, i.e. $N \neq \tilde{N}$. The reduced primal problem and the reduced dual problem yield a system of linear equations with dimensions N and $\tilde{N}$ respectively. For each time step and each parameter the solutions of the algebraic equations yield the primal solution coefficients $\{u_{N,n}^k(\xi)\}_{n=1}^N$ and the dual solution coefficients $\{\psi_{\tilde{N},n}^k(\xi)\}_{n=1}^{\tilde{N}}$. They determine a unique representation of the reduced solutions, for $k = 0, \dots, K$ and $\xi \in \Gamma$, such that

$$u_N^k(\xi) = \sum_{n=1}^{N} u_{N,n}^k(\xi)\varphi_n, \quad \psi_{\tilde{N}}^k(\xi) = \sum_{n=1}^{\tilde{N}} \psi_{\tilde{N},n}^k(\xi)\zeta_n.$$

The residual for the reduced primal problem is defined by, for $k = 1, \dots, K$,

$$r_N^k(v; \xi) = b(v; \xi) - \frac{1}{\Delta t}(u_N^k(\xi) - u_N^{k-1}(\xi), v)_{L^2(\Omega)} - a(u_N^k(\xi), v; \xi), \quad \forall v \in X_h.$$

The residual for the reduced dual problem is defined by, for $k = 0, \ldots, K-1$,

$$\tilde{r}^k_{\tilde{N}}(v;\xi) = -\frac{1}{\Delta t}(v, \psi^k_{\tilde{N}}(\xi) - \psi^{k+1}_{\tilde{N}}(\xi))_{L^2(\Omega)} - a(v, \psi^k_{\tilde{N}}(\xi);\xi), \qquad \forall v \in X_h.$$

Note, the primal residual and the dual residual are orthogonal onto their reduced spaces, i.e. $X_N \subset \ker(r^k_N(\cdot;\xi))$ and $\widetilde{X}_{\tilde{N}} \subset \ker(\tilde{r}^k_{\tilde{N}}(\cdot;\xi))$. In addition, the residual of the final condition (13) is given by,

$$\tilde{r}_{\mathrm{fc}}(v) = l(v) - (v, \psi^K_{\tilde{N}})_{L^2(\Omega)}, \quad \forall v \in X_h. \tag{14}$$

The reduced output is determined by

$$s_N(\xi) = l(u^K_N(\xi)) + \Delta t \sum_{k=1}^{K} r^k_N(\psi^{k-1}_{\tilde{N}}(\xi);\xi). \tag{15}$$

The second term in (15) is a "correction term", that uses the reduced dual solutions $\{\psi^k_{\tilde{N}}\}_{k=0}^{K-1}$ in order to achieve higher accuracy for the output computation. Typically the "correction" doubles the order of accuracy for the output approximation, cf. [16].

For computational efficiency, the computation is split into an offline phase and an online phase. The former is expensive to compute and depends on the large dimension $\mathcal{N}$, and it is related to the reduced model construction. Once the reduced model exists, solutions are obtained very fast in the online phase by calculations depending only on the reduced dimension N. For such a splitting, the bilinear form $a(\cdot,\cdot;\xi)$ and the functional $b(\cdot;\xi)$ need to be affine with respect to ξ (also known as parameter separable), i.e.

$$a(v,w;\xi) = \sum_{q=1}^{Q_a} \theta^a_q(\xi) a_q(v,w), \quad \forall v,w \in X, \forall \xi \in \Gamma, \tag{16}$$

$$b(v;\xi) = \sum_{q=1}^{Q_b} \theta^b_q(\xi) b_q(v), \qquad \forall v \in X, \forall \xi \in \Gamma. \tag{17}$$

If this assumption does not hold, an empirical interpolation method (EIM) [1] can be used instead.

4 Error Estimation

The objective in this section is the a posteriori error estimation in order to assess the accuracy of the reduced models. Meaning, the error between the detailed output and the reduced output is measured in a given norm. Rigorous error bounds for the solution error and the output error will be stated. By the assumptions for an offline-online decomposition in (16) and (17), the error bounds are computationally inexpensive, compared to the exact error computation. The error bounds are computed by the dual norm of the residuals. The following error estimators can be extended for a non-symmetric bilinear form $a(\cdot, \cdot; \xi)$.

4.1 Non-weighted Error Estimators

The following statements are taken from [6]: For the final condition (13), the error can be estimated by,

$$\left\| \psi_h^K - \psi_{\tilde{N}}^K \right\|_{L^2(\Omega)} \leq \Delta_{\tilde{N}}^{\psi,\mathrm{fc}} := \sup_{v \in X_h} \frac{|\tilde{r}_{\mathrm{fc}}(v)|}{\|v\|_{L^2(\Omega)}}, \tag{18}$$

where the final condition residual (14) is maximized. Note, if the reduced space for the dual problem contains $\psi_h^K \in \widetilde{X}_{\tilde{N}}$, the error estimator of the final condition is zero, i.e. $\Delta_{\tilde{N}}^{\psi,\mathrm{fc}} = 0$. The solution errors of the primal problem and the dual problem are estimated in parameter-dependent energy norms,

$$|||v|||_\xi^{\mathrm{pr}} := \left(\left\| v^K \right\|_{L^2(\Omega)}^2 + \Delta t \sum_{k=1}^{K} \left\| v^k \right\|_\xi^2 \right)^{1/2}, \qquad \forall v \in X_h^K,$$

$$|||v|||_\xi^{\mathrm{du}} := \left(\left\| v^0 \right\|_{L^2(\Omega)}^2 + \Delta t \sum_{k=0}^{K-1} \left\| v^k \right\|_\xi^2 \right)^{1/2}, \qquad \forall v \in X_h^K,$$

where $\|\cdot\|_\xi$ is defined in (3). Rigorous error estimators for the primal solution, dual solution and output for all $\xi \in \Gamma$, are defined by,

$$|||u_h(\xi) - u_N(\xi)|||_\xi^{\mathrm{pr}} \leq \Delta_N^u(\xi) := \left(\frac{\Delta t}{\alpha(\xi)} \sum_{k=1}^{K} \left\| r_N^k(\cdot; \xi) \right\|_{X'}^2 \right)^{1/2}, \tag{19}$$

$$|||\psi_h(\xi) - \psi_{\tilde{N}}(\xi)|||_\xi^{\mathrm{du}} \leq \Delta_{\tilde{N}}^\psi(\xi) := \left(\frac{\Delta t}{\alpha(\xi)} \sum_{k=0}^{K-1} \left\| \tilde{r}_{\tilde{N}}^k(\cdot; \xi) \right\|_{X'}^2 + (\Delta_{\tilde{N}}^{\psi,\mathrm{fc}})^2 \right)^{1/2}, \tag{20}$$

$$|s_h(\xi) - s_N(\xi)| \leq \Delta_{N,\tilde{N}}^s(\xi) := \Delta_N^u(\xi) \Delta_{\tilde{N}}^\psi(\xi). \tag{21}$$

For the error estimators the coercivity constant $\alpha(\xi)$, defined in (1), comes in. It can be approximated by a successive constraint method (SCM) [11] for instance. The dual norms of the residuals in (19) and (20) are computed by means of the Riesz representation theorem (see, e.g., [2, Chapter 2.4]).

4.2 Weighted Error Estimators

The pdf $\rho(\cdot)$ appears in the computation of statistical quantities. Using the results from the previous section, the expected solution error can be estimated by,

$$\mathbb{E}[|||u_h - u_N|||_\xi^{\text{pr}}] = \int_\Gamma |||u_h(\xi) - u_N(\xi)|||_\xi^{\text{pr}} \rho(\xi)\mathrm{d}\xi \leq \int_\Gamma \Delta_N^u(\xi)\rho(\xi)\mathrm{d}\xi,$$

just as the expected output error

$$\mathbb{E}[|s_h - s_N|] = \int_\Gamma |s_h(\xi) - s_N(\xi)|\rho(\xi)\mathrm{d}\xi \leq \int_\Gamma \Delta_{N,\tilde{N}}^s(\xi)\rho(\xi)\mathrm{d}\xi.$$

The following weighted error estimators, introduced in [3] for elliptic problems, are defined,

$$\Delta_N^{u,\rho}(\xi) := \Delta_N^u(\xi)\rho(\xi), \qquad \forall \xi \in \Gamma, \tag{22}$$

$$\Delta_{N,\tilde{N}}^{s,\rho}(\xi) := \Delta_{N,\tilde{N}}^s(\xi)\rho(\xi), \quad \forall \xi \in \Gamma, \tag{23}$$

where the weight $\rho(\xi)$ gives greater weight to more likely parameter values. The weighted estimators (22) and (23) are still computationally cheap and will be used as a optimality criterion for a weighted reduced space construction.

5 Reduced Space Construction

In this section, a non-weighted and a weighted reduced space construction are described. The reduced space construction is based on the POD-greedy algorithm [9], stated in Algorithm 1.

Data: ϵ_{tol}, parameter training set $\Gamma_{\text{train}} \subset \Gamma$, $\xi^{(1)}$
Result: reduced space X_N
$N = 1,\ X_1 = \text{span}\{u_h^K(\xi^{(1)})\}$
while $\epsilon_N := \max_{\xi\in\Gamma_{train}} \Delta_N(\xi) > \epsilon_{tol}$ **do**

$\xi^{(N+1)} := \arg\max_{\xi\in\Gamma_{\text{train}}} \Delta_N(\xi)$

compute $u_h^k(\xi^{(N+1)}),\ k = 0, \ldots, K$, using (10) and (11)

$e_P^k(\xi^{(N+1)}) := u_h^k(\xi^{(N+1)}) - P_{X_N} u_h^k(\xi^{(N+1)}),\ k = 0, \ldots, K$

$\varphi_{N+1} := POD_1(\{e_P^k(\xi^{(N+1)})\}_{k=0}^K)$

$X_{N+1} := X_N \oplus \text{span}\{\varphi_{N+1}\}$

$N := N + 1$

end

Algorithm 1: POD-greedy algorithm

Note, the algorithm is formulated for the primal problem (6), (7), (10), and (11). Analogously the algorithm can be utilized for the dual problem (8), (9), (12), and (13).

The POD-greedy algorithm combines a greedy algorithm for the parameter domain and a POD for the time interval. It computes the detailed solution $u_h^K(\xi^{(1)})$, for a given parameter $\xi^{(1)}$, which spans the initial reduced space X_1. The dimension of the reduced basis N grows iteratively. As a stopping criterion for the iteration, an error estimator of the previous section needs to fall below some given error tolerance $\epsilon_{\text{tol}} > 0$. Alternatively, a predefined reduced dimension can be given as a stopping criterion. The parameter value ξ^{N+1} is determined in each iteration by evaluating an optimality criterion. Maximizing the exact errors over a large parameter domain can be computationally infeasible. Instead, the computationally cheap error estimators are maximized, which is often called weak POD-greedy. The maximum is sought over a training parameter set, which is a finite and uniformly sampled approximation set of the parameter domain. The solutions for the single time steps $\{u_h^k(\xi^{(N+1)})\}_{k=1}^K$, called snapshots, are computed. In order to compress the information of the obtained solution trajectory, the first POD mode of the projection error onto the reduced space is computed and added to the reduced basis.

5.1 A Non-Weighted Reduced Space Construction

A non-weighted reduced space construction is based on Algorithm 1. It is distinguished if the objective is either the approximation of the solution u_h or the approximation of the output s_h. The former uses the primal solution error estimator in (19) for the optimality criterion in the POD-greedy procedure. The latter uses the output error estimator in (21) for the optimality criterion in the POD-greedy procedure. The output error estimator consists of a primal error estimator (19) and dual error estimator (20). Therefore, maximizing the output error estimator yields a reduced space regarding the primal problem and a reduced space regarding the dual problem. Note, a non-weighted reduced space construction weights all parameter values $\xi \in \Gamma$ equally.

5.2 A Weighted Reduced Space Construction

As in the previous section, a weighted reduced space construction is based on Algorithm 1. However, in this section the objective is to build up a reduced space, that gives better error convergence rates regarding statistical quantities, compared to the non-weighted approach. Since the input parameters are random, certain parameters are in general more likely to appear. Highly probable parameters obtain more importance incorporating the pdf. Hence, weighted error estimators of Sect. 4.2 are used for the reduced space construction. It is distinguished if the objective is either the approximation of the expected solution $\mathbb{E}[u_h]$ or the approximation of the expected output $\mathbb{E}[s_h]$. The former uses the weighted primal solution error estimator in (22) for the optimality criterion in the POD-greedy procedure. The latter uses the weighted output error estimator in (23) for the optimality criterion in the POD-greedy procedure. The weighted output error estimator consists of a primal error estimator (19) and dual error estimator (20). Therefore, maximizing the weighted output error estimator yields a weighted reduced space regarding the primal problem and a weighted reduced space regarding the dual problem. Note, the weighted approach uses the same uniformly sampled parameter set Γ_{train} as the non-weighted approach. Meaning, the weighting comes in only by the optimality criterion that maximizes the weighted error estimators.

6 ROM Comparison with a POD Projection

In this work, a conceptual comparison between different ROMs is drawn. In Sects. 5.1 and 5.2 a non-weighted and a weighted ROM construction by a POD-greedy approach were shown. In this section, the idea is to build up another reduced basis that yields an optimal reduced solution regarding the expected solution error, namely

$$\min_{w_1,\dots,w_N\in X_h} \quad \min_{\substack{u^1,\dots,u^K:\\ \Gamma\to\operatorname{span}\{w_1,\dots,w_N\}}} \quad \mathbb{E}\left[\Delta t\sum_{k=1}^{K}\left\|u_h^k-u^k\right\|_{\xi_{\text{ref}}}^2\right], \tag{24}$$
$$\text{subject to } (w_m,w_n)_{\xi_{\text{ref}}}=\delta_{mn}.$$

The norm $\|\cdot\|_{\xi_{\text{ref}}}$ and the inner product $(\cdot,\cdot)_{\xi_{\text{ref}}}$ are defined by (4) and δ_{mn} is the Kronecker delta. An optimal choice for the unknown functions $\{u^k\}_{k=1}^K$ can be achieved by choosing their orthogonal projection onto the N-dimensional POD space $X_{\text{POD},N}:=\operatorname{span}\{w_1,\dots,w_N\}$, namely

$$u_{\text{POD},N}^k(\xi):=\sum_{n=1}^{N}(u_h^k(\xi),w_n)_{\xi_{\text{ref}}}w_n\in X_{\text{POD},N},\quad k=1,\dots K. \tag{25}$$

The orthogonal projection yields an equivalent formulation of the minimization problem (24), such that

$$\min_{w_1,\dots,w_N\in X_h}\ \mathbb{E}\left[\Delta t\sum_{k=1}^{K}\left\|u_h^k-\sum_{n=1}^{N}(u_h^k,w_n)_{\xi_{\text{ref}}}w_n\right\|_{\xi_{\text{ref}}}^2\right], \tag{26}$$
$$\text{subject to } (w_m,w_n)_{\xi_{\text{ref}}}=\delta_{mn}.$$

The expectation in (24) and (26) can not be determined analytically. Hence, the expectation is approximated by a Monte Carlo (MC) method (see, e.g., [5, Chapter 2.7.3]). It uses snapshots $\{u_h^k(\xi^{(i)})\}_{i=1}^{N_{\mathrm{MC}}}$, with random realizations $\{\xi^{(i)}\}_{i=1}^{N_{\mathrm{MC}}}$ sampled by its pdf, such that

$$\mathbb{E}_{\mathrm{MC}}[u_h^k] = \frac{1}{N_{\mathrm{MC}}} \sum_{i=1}^{N_{\mathrm{MC}}} u_h^k(\xi^{(i)}), \quad k = 1, \ldots, K. \tag{27}$$

Note, the number of MC samples N_{MC} is assumed to be large enough, such that the MC error can be neglected, i.e. $\mathbb{E}[\cdot] \approx \mathbb{E}_{\mathrm{MC}}[\cdot]$. By the MC approximation of the expectation the minimization problem in (26) is given by

$$\min_{w_1,\ldots,w_N \in X_h} \frac{\Delta t}{N_{\mathrm{MC}}} \sum_{i=1}^{N_{\mathrm{MC}}} \sum_{k=1}^{K} \left\| u_h^k(\xi^{(i)}) - \sum_{n=1}^{N} (u_h^k(\xi^{(i)}), w_n)_{\xi_{\mathrm{ref}}} w_n \right\|_{\xi_{\mathrm{ref}}}^2, \tag{28}$$
$$\text{subject to } (w_m, w_n)_{\xi_{\mathrm{ref}}} = \delta_{mn}.$$

This minimization problem can be solved by an eigenvalue problem, see, e.g. [7, Theorem 1.8]. The resulting eigenfunctions $\{w_n\}_{n=1}^N$ are the first $N \in \{1, \ldots, N_{\mathrm{MC}} K\}$ orthogonal POD modes and the eigenvalues $\{\lambda_l\}_{l=1}^{N_{\mathrm{MC}} K}$ determine the error in (28), such that

$$\frac{\Delta t}{N_{\mathrm{MC}}} \sum_{i=1}^{N_{\mathrm{MC}}} \sum_{k=1}^{K} \left\| u_h^k(\xi^{(i)}) - \sum_{n=1}^{N} (u_h^k(\xi^{(i)}), w_n)_{\xi_{\mathrm{ref}}} w_n \right\|_{\xi_{\mathrm{ref}}}^2 = \sum_{l=N+1}^{N_{\mathrm{MC}} K} \lambda_l.$$

The first N POD modes span the POD space $X_{\mathrm{POD},N} = \mathrm{span}\{w_1, \ldots, w_N\}$.

However, the optimality pays its price, since the POD needs the detailed solution for each parameter in the parameter domain. Compared to the POD-greedy in Sect. 5, the evaluation of the error estimator for the parameter values are computationally cheap and eventually only N detailed solutions for the reduced space construction need to be evaluated. Note, the comparison only can be done for the expected solution error. For the expected output error $\mathbb{E}[|s_h - s_N|]$ an optimal reduced space cannot be found with a POD.

In this section optimality for the mean square error could be stated. Therefore, the POD method determines an optimal reduced POD space based on a finite set of snapshots and a parameter-independent energy norm. Moreover, the optimal representation of the reduced solution is given by the orthogonal projection onto the reduced POD space. The minimal mean square error can be easily computed by the truncated eigenvalues coming out of the POD method.

7 Numerical Example

An instationary heat transfer is considered [6] as a numerical example. The time dependent heat flow is computed on a rectangular domain where three squares are cut out, see Fig. 1. The spatial domain is defined by $\Omega = \{[0, 10] \times [0, 4]\} \setminus \{\{(1, 3) \times (1, 3)\} \cup \{(4, 6) \times (1, 3)\} \cup \{(7, 9) \times (1, 3)\}\}$. As a quantity of interest $s(\cdot)$, the average temperature in the domain at the end time point T is computed. The stochastic parameters $\xi = (\xi^{\text{out}}, \xi^{\text{in}})$ enter the problem via the boundary condition. On the left domain boundary $\partial\Omega_{\text{out}}$ there is a random heat inflow $\kappa(\cdot; \xi^{\text{out}})$ modeled by a Karhunen-Loève (KL) expansion [13, Chapter 37.5]. Therefore, it needs to hold that κ is a second-order random field, meaning that its second moment is finite. As a consequence of the KL, the corresponding random variables ξ^{out} are uncorrelated and have zero mean. The top, right and bottom boundary are insulated. The condition at the inner boundary of the domain (squares) $\partial\Omega_{\text{in}}$ is parametrized by a beta distributed parameter ξ^{in}, that can be interpreted as a cooling parameter.

The weak formulation of the parabolic problem reads as follows: For given realization $\xi = \left(\xi^{\text{out}}, \xi^{\text{in}}\right) \in \Gamma$, evaluate

$$s(\xi) = \frac{1}{|\Omega|} \int_{\Omega} u(T; \xi),$$

where for any $t \in [0, T]$ the solution $u(t; \xi) \in X := H^1(\Omega)$ fulfills

$$\int_{\Omega} \partial_t u v + \int_{\Omega} \nabla u \cdot \nabla v + \int_{\partial\Omega_{\text{out}}} \kappa(\xi^{\text{out}}) u v + \xi^{\text{in}} \int_{\partial\Omega_{\text{in}}} u v = \int_{\partial\Omega_{\text{out}}} \kappa(\xi^{\text{out}}) v,$$
$$u(0; \xi) = 0,$$

for all $v \in X$. The boundary integrals result from a parametrized Robin boundary condition,

$$\frac{\partial u}{\partial n} = \begin{cases} \kappa(\xi^{\text{out}})(1 - u) & \text{on } \partial\Omega_{\text{out}}, \\ -\xi^{\text{in}} u & \text{on } \partial\Omega_{\text{in}}, \\ 0 & \text{on } \partial\Omega \setminus \{\partial\Omega_{\text{out}} \cup \partial\Omega_{\text{in}}\}. \end{cases}$$

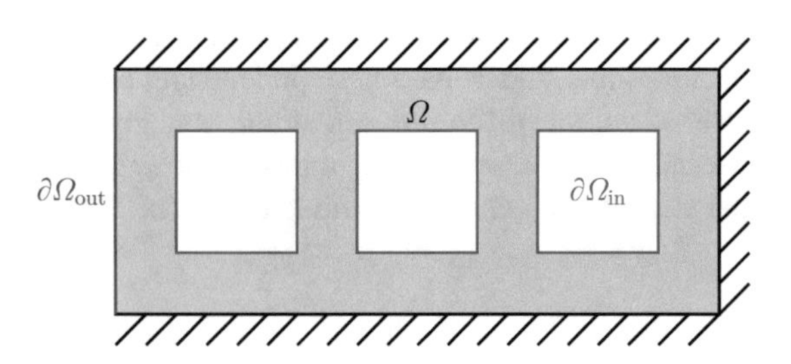

Fig. 1 Heat flow in domain Ω is considered

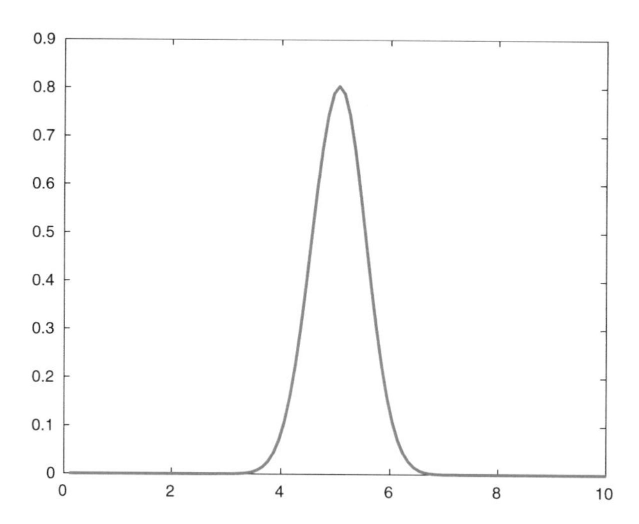

Fig. 2 Pdf of beta distribution with support [0.1, 10] and scaling parameter 50

The time interval is determined by $T = 20$, and the KL field is given by its truncated version $\kappa(x, \xi^{\text{out}}) = \alpha_0(x) + \sum_{l=1}^{N_{\text{KL}}} \sqrt{\mu_l} c_l(x) \xi_l^{\text{out}}$ where the first $N_{\text{KL}} = 10$ eigenpairs are chosen. The eigenvalues $\{\mu_l\}_{l=1}^{N_{\text{KL}}}$ and the eigenfunctions $\{c_l\}_{l=1}^{N_{\text{KL}}}$ are solutions of an integral equation containing a covariance operator [14, Chapter 7.4], corresponding to an exponential kernel $C(x, y) = \exp(-|x - y|/a)$ with correlation length $a = 2$. The expectation of the random field is defined by $\alpha_0 = 10$. The random parameters $\xi^{\text{out}} = \{\xi_l^{\text{out}}\}_{l=1}^{N_{\text{KL}}}$ are assumed to be independent of each other and are distributed with uniform distribution $\mathcal{U}\left(-\sqrt{3}, \sqrt{3}\right)$. The random parameter ξ^{in} is distributed with beta distribution $\mathcal{B}\,(0.1, 10, 50, 50)$ where the first two inputs are the interval bounds of the support, the third and fourth input are scaling parameters. In Fig. 2 its probability density function is drawn. Further, the parameters ξ^{in} and ξ^{out} are independent of each other as well.

The bilinear form, right hand side and linear output functional are defined by,

$$a(w, v; \xi) := \int_{\Omega} \nabla w \cdot \nabla v + \int_{\partial\Omega_{\text{out}}} \kappa(\xi^{\text{out}}) w v + \xi^{\text{in}} \int_{\partial\Omega_{\text{in}}} w v, \quad \forall w, v \in X,$$

$$b(v; \xi) := \int_{\partial\Omega_{\text{out}}} \kappa(\xi^{\text{out}}) v, \quad l(v) := \frac{1}{|\Omega|} \int_{\Omega} v, \quad \forall v \in X.$$

For the spatial discretization, a linear finite element method with $\mathcal{N} = 1132$ degrees of freedom is used. For the time discretization, an implicit Euler method is applied with time step size $\Delta t = 0.2$ and $K = 100$. In Fig. 3, two solutions can be seen, the finite element solution for the reference parameter $\xi_{\text{ref}} = (0, \ldots, 0, 0.1) \in \Gamma$ in

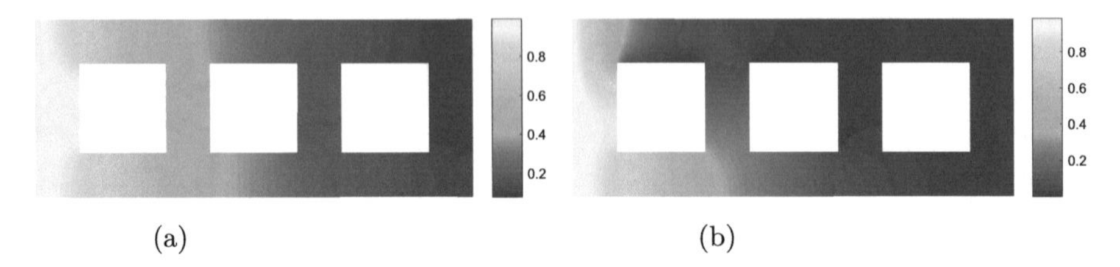

Fig. 3 Finite element solutions at time point T. (**a**) Reference solution. (**b**) Random solution

Fig. 3a and the finite element solution for a randomly sampled parameter in Fig. 3b. Recall, the reference parameter defines an inner product, see (4), such that

$$(w, v)_{\xi_{\text{ref}}} = \int_{\Omega} \nabla w \cdot \nabla v + 10 \int_{\partial \Omega_{\text{out}}} wv + 0.1 \int_{\partial \Omega_{\text{in}}} wv, \quad \forall w, v \in X.$$

The coercivity constant in (1) is chosen as a uniformly lower bound, i.e. $\overline{\alpha} = 1$. For the reduced space construction, see Sect. 5, the parameter domain $\Gamma = \left[-\sqrt{3}, \sqrt{3}\right]^{N_{\text{KL}}} \times [0.1, 10]$ is approximated by a subset $\Gamma_{\text{train}} \subset \Gamma$, which contains $|\Gamma_{\text{train}}| = 500$ independent uniformly distributed parameter samples. All the reduced spaces regarding the primal problem have the same initial basis. As an initial parameter value $\xi^{(1)}$, the parameter is chosen such that ξ^{in} attains its maximum in Γ_{train}. The solution at the end time point for that parameter spans the initial reduced basis, e.g. $X_1 = \text{span}\{u_h^K(\xi^{(1)})\}$. Further, all the reduced spaces regarding the dual problem have the same initial basis. It simply takes the solution of the final condition as initial basis, e.g. $\widetilde{X}_1 = \text{span}\{\psi_h^K\}$. This choice implies that the error estimator for the final condition in (18) is zero.

In the following different ROMs are compared. Therefore, the root mean square error and the mean absolute output error are considered.

First, the goal is an efficient approximation of the root mean square solution error. Therefore, three different models are considered. The first reduced space X_N^u, is obtained by a non-weighted POD-greedy algorithm, see Algorithm 1. It uses a non-weighted error estimator (19) as the optimality criterion. The non-weighted reduced solutions $\{u_N^k \in X_N^u\}_{k=1}^K$ are determined by (10) and (11). The second reduced space $X_N^{u,\rho}$, is obtained by a weighted POD-greedy algorithm. It uses a weighted error estimator (22) as the optimality criterion in Algorithm 1. The weighted reduced solutions $\{u_{N,\rho}^k \in X_N^{u,\rho}\}_{k=1}^K$ are determined by (10) and (11). The third model is obtained by a POD, see Sect. 6. Therefore, the snapshots $\{u_h^k(\xi^{(i)})\}_{i,k=1}^{N_{\text{MC}},K}$ for each time step and for all parameter values, sampled by the joint pdf, are computed. The POD basis functions determine the POD solutions $\{u_{\text{POD},N}^k \in X_{\text{POD},N}\}_{k=1}^K$, see (25). Those three solutions determine the expected solution errors in Fig. 4a, where the errors for the first 30 basis functions are shown. For the approximation of the expected value, the MC method (27) is used. Therefore, the same samples as for the POD snapshot space are chosen.

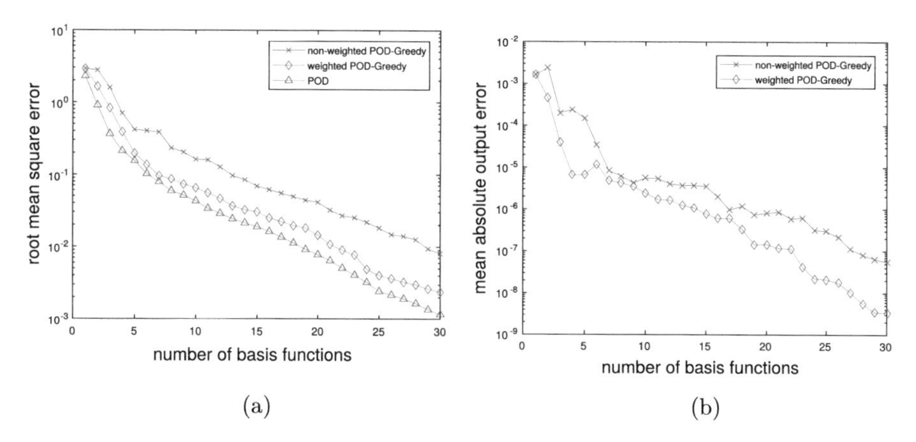

Fig. 4 Comparison between non-weighted, weighted POD-greedy and POD. (**a**) $\mathbb{E}_{\mathrm{MC}}\left[\Delta t \sum_{k=1}^{K} \|u_h^k - u_N^k\|_{\xi_{\mathrm{ref}}}^2\right]^{1/2}$. (**b**) $\mathbb{E}_{\mathrm{MC}}[|s_h - s_N|]$

Second, the goal is an efficient approximation of the mean absolute output error, where a non-weighted and a weighted approach are compared. The former utilizes a non-weighted POD-greedy algorithm, that uses a non-weighted output error estimator (21) as the optimality criterion. The output error estimator consists of an error estimator for the primal solution (19) and an error estimator for the dual solution (20). As explained at the end of Sect. 5.1, the non-weighted POD-greedy for the output approximation yields two reduced spaces X_N^s and $\widetilde{X}_{\tilde{N}}^s$. The non-weighted reduced output is determined by (15). The calculation of the reduced output is based on the reduced solutions $u_N^K \in X_N^s$, computed by (10) and (11), and $\{\psi_{\tilde{N}}^k \in \widetilde{X}_{\tilde{N}}^s\}_{k=0}^{K-1}$, computed by (12) and (13). The weighted approach utilizes a weighted POD-greedy algorithm, that uses a weighted output error estimator (23) as the optimality criterion. The weighted output error estimator consists of an error estimator for the primal solution (19) and an error estimator for the dual solution (20). As explained at the end of Sect. 5.2, the weighted POD-greedy for the output approximation yields two reduced spaces $X_N^{s,\rho}$ and $\widetilde{X}_{\tilde{N}}^{s,\rho}$. The weighted reduced output is determined by (15). The calculation of the reduced output is based on the reduced solutions $u_N^K \in X_N^{s,\rho}$, computed by (10) and (11), and $\{\psi_{\tilde{N}}^k \in \widetilde{X}_{\tilde{N}}^{s,\rho}\}_{k=0}^{K-1}$, computed by (12) and (13). Those two approaches determine the mean absolute output errors in Fig. 4b, where the errors for the first 30 basis functions are shown. Figure 4 shows that in each iteration of the POD-greedy the weighted approach gives better error results compared to the non-weighted approach. Further, the optimal error convergence, obtained from the POD, is observed in Fig. 4a.

8 Conclusions

In this work, different model order reduction techniques for a specific parabolic model problem with data uncertainties were studied. Namely, a RBM, a weighted RBM and a POD were used in order to reduce the dimension of a high-fidelity model obtained from a linear finite element model. Apart from the solution, a linear functional maps the solution to a quantity of interest. The work considered the root mean square error and the mean absolute output error. A numerical example for an instationary heat transfer with random input data was studied. It has been shown, that a weighted RBM yields better error results of the root mean square error and the mean absolute output error compared to a non-weighted RBM. The POD yields an optimal reduced space of the mean square error for an energy norm. However, the reduced space construction using a POD requires the snapshots over the parameter domain differently from the RBM. Hence, for a large parameter set this can be computationally challenging. The RBM utilizes the computationally cheap error estimators for the reduced space construction instead. The errors for the numerical example show, that the weighted RBM is closer to the POD than the non-weighted RBM.

Acknowledgements This work is supported by the 'Excellence Initiative' of the German federal and state governments and the Graduate School of Computational Engineering at the Technische Universität Darmstadt.

References

1. Barrault, M., Maday, Y., Nguyen, N.C., Patera, A.T.: An 'empirical interpolation' method: application to efficient reduced-basis discretization of partial differential equations. C. R. Math. **339**(9), 667–672 (2004). https://doi.org/10.1016/j.crma.2004.08.006
2. Brenner, S.C., Scott, L.R.: The Mathematical Theory of Finite Element Methods, 3rd edn. Springer, New York (2008)
3. Chen, P., Quarteroni, A., Rozza, G.: A weighted reduced basis method for elliptic partial differential equations with random input data. SIAM J. Numer. Anal. **51**(6), 3163–3185 (2013). https://doi.org/10.1137/130905253
4. Chen, P., Quarteroni, A., Rozza, G.: Reduced basis methods for uncertainty quantification. SIAM/ASA J. Uncertain. Quantif. **5**(1), 813–869 (2017). https://doi.org/10.1137/151004550
5. Fishman, G.: Monte Carlo: Concepts, Algorithms, and Applications. Springer, New York (1996)
6. Grepl, M.A., Patera, A.T.: A posteriori error bounds for reduced-basis approximations of parametrized parabolic partial differential equations. ESAIM Math. Model. Numer. Anal. **39**(1), 157–181 (2005). https://doi.org/10.1051/m2an:2005006
7. Gubisch, M., Volkwein, S.: Proper orthogonal decomposition for linear-quadratic optimal control. In: Benner, P., Ohlberger, M., Cohen, A., Willcox, K. (eds.) Reduction and Approximation: Theory and Algorithms, Chap. 1. Society for Industrial and Applied Mathematics, Philadelphia (2017). https://doi.org/10.1137/1.9781611974829.ch1

8. Haasdonk, B.: Reduced basis methods for parametrized pdes– a tutorial introduction for stationary and instationary problems. In: Benner, P., Ohlberger, M., Cohen, A., Willcox, K. (eds.) Reduction and Approximation: Theory and Algorithms, Chap. 2. Society for Industrial and Applied Mathematics, Philadelphia (2017). https://doi.org/10.1137/1.9781611974829.ch2
9. Haasdonk, B., Ohlberger, M.: Reduced basis method for finite volume approximations of parametrized linear evolution equations. ESAIM Math. Model. Nume. Anal. **42**(2), 277–302 (2008). https://doi.org/10.1051/m2an:2008001
10. Hesthaven, J.S., Rozza, G., Stamm, B.: Certified Reduced Basis Methods for Parametrized Partial Differential Equations. Springer International Publishing, New York (2016). https://doi.org/10.1007/978-3-319-22470-1
11. Huynh, D.B.P., Rozza, G., Sen, S., Patera, A.T.: A successive constraint linear optimization method for lower bounds of parametric coercivity and inf—sup stability constants. C. R. Math. **345**(8), 473–478 (2007). https://doi.org/10.1016/j.crma.2007.09.019
12. Kunisch, K., Volkwein, S.: Galerkin proper orthogonal decomposition methods for a general equation in fluid dynamics. SIAM J. Numer. Anal. **40**(2), 492–515 (2002). https://doi.org/10.1137/S0036142900382612
13. Loève, M.: Probability Theory II, 4th edn. Springer, New York (1978)
14. Lord, G.J., Powell, C.E., Shardlow, T.: An Introduction to Computational Stochastic PDEs. Cambridge University Press, Cambridge (2014)
15. Oden, J.T., Prudhomme, S.: Goal-oriented error estimation and adaptivity for the finite element method. Comput. Math. Appl. **42**(5–6), 735–756 (2001). https://doi.org/10.1016/S0898-1221(00)00317-5
16. Pierce, N.A., Giles, M.B.: Adjoint recovery of superconvergent functionals from pde approximations. SIAM Rev. **42**(2), 247–264 (2000). https://doi.org/10.1137/S0036144598349423
17. Prud'homme, C., Rovas, D.V., Veroy, K., Machiels, L., Maday, Y., Patera, A.T., Turinici, G.: Reliable real-time solution of parametrized partial differential equations: reduced-basis output bound methods. ASME. J. Fluids Eng. **124**(1), 70–80 (2001). https://doi.org/10.1115/1.1448332
18. Quarteroni, A., Manzoni, A., Negri, F.: Reduced Basis Methods for Partial Differential Equations an Introduction. Springer International Publishing, New York (2016). https://doi.org/10.1007/978-3-319-15431-2

Modeling of the Compressed-Air Flow Impact for Thermoforming Simulations

Simon Wagner, Manuel Münsch, Fabian Kayatz, Jens-Peter Majschak, and Antonio Delgado

Abstract Thermoforming is a process for the cheap mass-production of thin-walled plastic parts. A sheet of plastic is heated for increased deformability, and then deformed by overpressure into a mold with the end product's shape. The main drawback is the inhomogeneous wall thickness distribution resulting from the common process. The authors believe that it is possible to improve these inhomogeneities by locally influencing the highly temperature-dependent material strength using directed jets of pressurized air for local cooling. As the high number of potentially influential parameters renders purely experimental parameter studies infeasible, a computational model that couples the flow of the pressurized air with the structural simulation of the deforming plastic is set up. With the combined results of experiments and simulations, a parameter study can be conducted, which allows for an optimization of air flow parameters for a more evenly distributed wall thickness.

Keywords Thermoforming · Thermal fluid-structure-interaction

1 Motivation and State of the Art

Thermoforming is widely used in the production of thin-walled plastic mass products mainly used for packaging, like cups or blisters. Apart from an exact reproduction of the mold's shape, the end products must satisfy requirements that depend on the local wall thickness, e.g. stiffness or diffusion resistance properties [7]. In general, the more uniform the wall thickness distribution is,

S. Wagner (✉) · M. Münsch · A. Delgado
Friedrich-Alexander Universität Erlangen-Nürnberg, Lehrstuhl für Strömungsmechanik, Erlangen, Germany
e-mail: simon.sw.wagner@fau.de

F. Kayatz · J.-P. Majschak
Fraunhofer-Institut für Verfahrenstechnik und Verpackung IVV, Dresden, Germany

M. Schäfer et al. (eds.), *Recent Advances in Computational Engineering*, Lecture Notes in Computational Science and Engineering 124,
https://doi.org/10.1007/978-3-319-93891-2_10

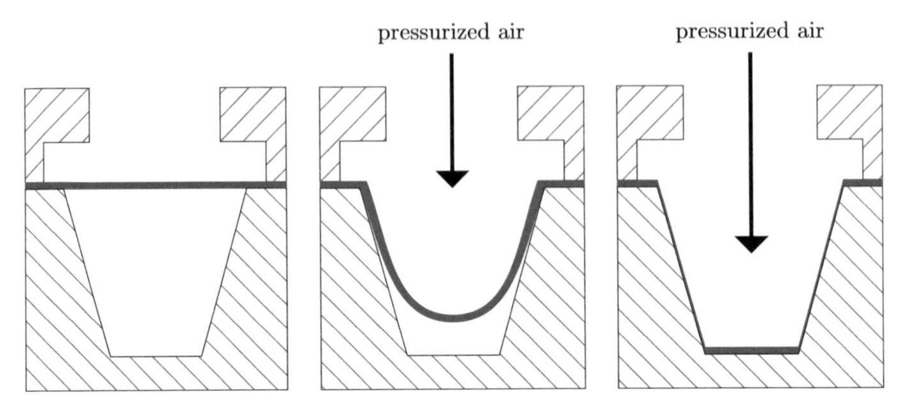

Fig. 1 Depiction of the common process without plug-assisted prestretching or directed air flow. The plastic sheet is shown in dark gray

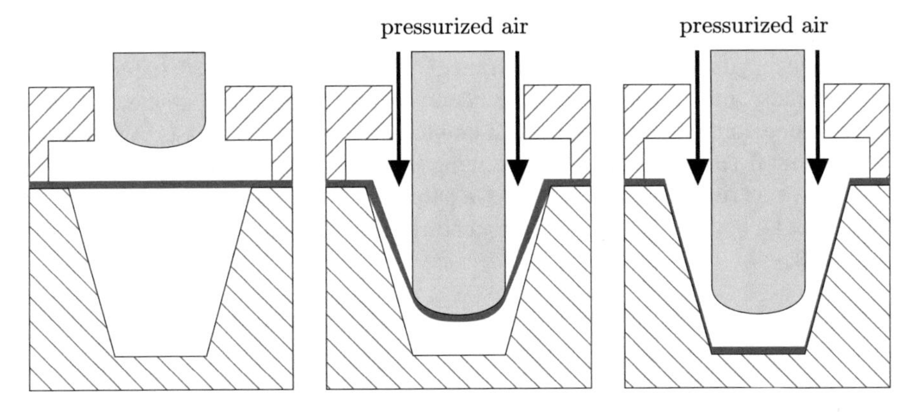

Fig. 2 Depiction of the common process with plug-assisted pre-stretching. The plastic sheet is shown in dark gray whereas the plug is shown in light gray. The pressurized air enters the pressure-box after the plug reached its maximum depth

the better these properties are fulfilled, which leads to the aim of a completely homogeneous wall thickness distribution.

In the common process (Fig. 1), the main source of inhomogeneity is supposed to be that different parts of the plastic sheet get in contact with the mold wall at different times [2]. It can be assumed that at wall contact, the heat transfer from the plastic to the mold wall leads to almost sudden cooling, so that further stretching happens mostly in the parts of the plastic sheet that did not yet touch the mold wall [6]. In the past years, research on improving the wall thickness distribution of thermoformed products focused mainly on plug pre-stretching techniques as shown in Fig. 2. This enhancement of the process allows for a certain reduction of the wall thickness inequalities and is used especially for the production of 'deep' packaging with high stretching ratios. Despite the advantage of more homogeneous wall thickness distributions, the overhead and more complicated machinery with additional parts constitute a drawback of this approach. Adaption of a plug to a new product may be time consuming, and the plugs only have limited durability. Furthermore, for the investigation and simulation of the process, the plug friction and heat transfer add substantial uncertainties, because both have a significant effect which is hard to measure [3, 6].

The authors believe that another possibility to positively influence the wall thickness distribution would be to indirectly influence the plastic sheet's material strength using the highly temperature-dependent material properties of common materials used for thermoforming like PE, PP, PET, or PVC. Semi-crystalline plastics in particular exhibit strong differences in strength already for small differences in temperature. It has already been shown that a local improvement in wall thickness can be achieved by directing air jets onto the plastic sheet [5]. By directing air jets, which increase local convective cooling, onto areas of the plastic sheets that require increased strength for an improved deformation behavior, the authors believe that it is possible to significantly improve the wall thickness distribution of the end product. Although the directed air flow also leads to a local pressure increase at the impingement location on the plastic sheet, its influence is way lower than what can be achieved by local cooling, so the thermal influence is considered to be significantly higher than the dynamic influence.

The authors set up a simulation model that includes only the influential effects in order to reduce the computational requirements, while maintaining enough accuracy to render representative simulations possible. In this regard, especially the qualitative features of the used numerical models are important, as experimental validation will allow for model calibration according to the experimental results.

Simulating the process outlined above, with the high mutual interdependency of the transient behavior of fluid and structural part, requires a tight coupling, including thermal effects. This tight coupling can with reasonable accuracy and computational demands only be achieved by an implicit thermal fluid-structure-interaction (TFSI) simulation.

2 Geometry of Test Molds

For the experiments, two test geometries have been chosen, that resemble close to commonly produced thermoformed cups. A notable difference are the relatively low corner blend radii of 1 mm, which have been chosen in order to stress the effect of thinning due to excessive stretching of the plastic sheet areas that are the last to touch the mold wall—which in the case of a simple cup is the lower circumferential corner. The cross section of the first and axis-symmetric test mold is shown in Fig. 3. For the given dimensioning the radius is $R = 22.5$ mm, the depth is $H = 25$ mm and the wall slope is 3°. At the mold's bottom, there are venting holes, that are necessary for the escape of the air that fills the mold at the start of the process. For the first mold, the mean stretching ratio, which is the ratio of the areas in final and initial

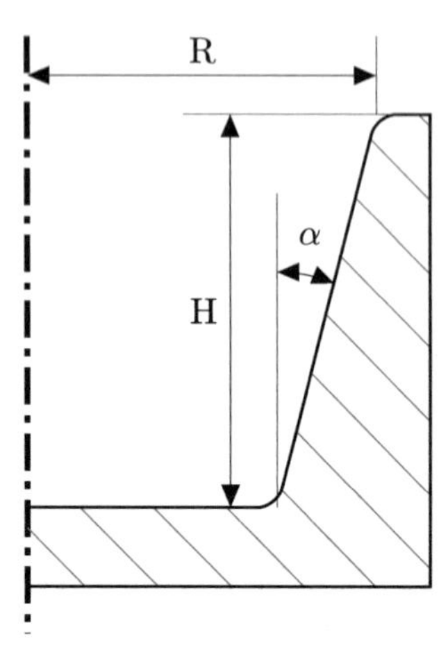

Fig. 3 Cross sectional view of the mold. Venting holes at the mold's bottom are not shown

configuration, is 2.82. This means that at the end of the thermoforming process, one can expect a mean sheet thickness of 35.4% of the initial thickness. The second test mold will be subject to later experiments and is not shown. It is cuboid-like with two axes of symmetry, a length of 56 mm, a width of 28 mm and a depth of 20.7 mm, which implies an approximate mean stretching ratio of 2.76.

3 Simulation Setup

A cross-sectional sketch of the three-dimensional simulation domain, both for the initial and the final state of the thermoforming process, can be seen in Fig. 4. Whereas in the real process the plastic sheet extends further outwards, where it is clamped between the pressure-box and the mold, the simulated part is restricted to the region that is surrounded by air on both sides of the sheet. The high deformation of the plastic sheet causes a significant enlargement of the fluid domain during the simulation, which in turn makes a remeshing of the fluid domain during simulation run indispensable. There are only few computational fluid dynamics solvers with the capabilities of automated remeshing that also provide an interface for automated thermally coupled fluid-structure-interaction. Therefore, for the current simulations, ANSYS Fluent has been chosen in combination with ANSYS Mechanical for the structural simulation part.

While it would have been entirely acceptable to reduce the structural part to a quarter of the domain by exploiting the domain's double mirror symmetry, for the fluid part the symmetry boundary condition would have implied unphysical restrictions on the turbulence field, because turbulence is an inherently three-dimensional phenomenon. Therefore the simulations were carried out using a completely three-dimensional domain.

For further studies, the second, non-axis-symmetric test mold will also be investigated in simulations, but the current studies focus on the axis-symmetric mold form.

3.1 Fluid Simulation

The fluid domain is simulated using ANSYS Fluent, a general-purpose finite-volume-based computational fluid dynamics code. There are two fluid zones (shown hatched in Fig. 4): one of which has fixed boundaries and therefore a fixed computational mesh, the other one, with the moving interface as a boundary requires a mesh which is dynamically adapted and locally remeshed during the simulation. For both zones tetrahedral cells are used, which allows for simple refining in the

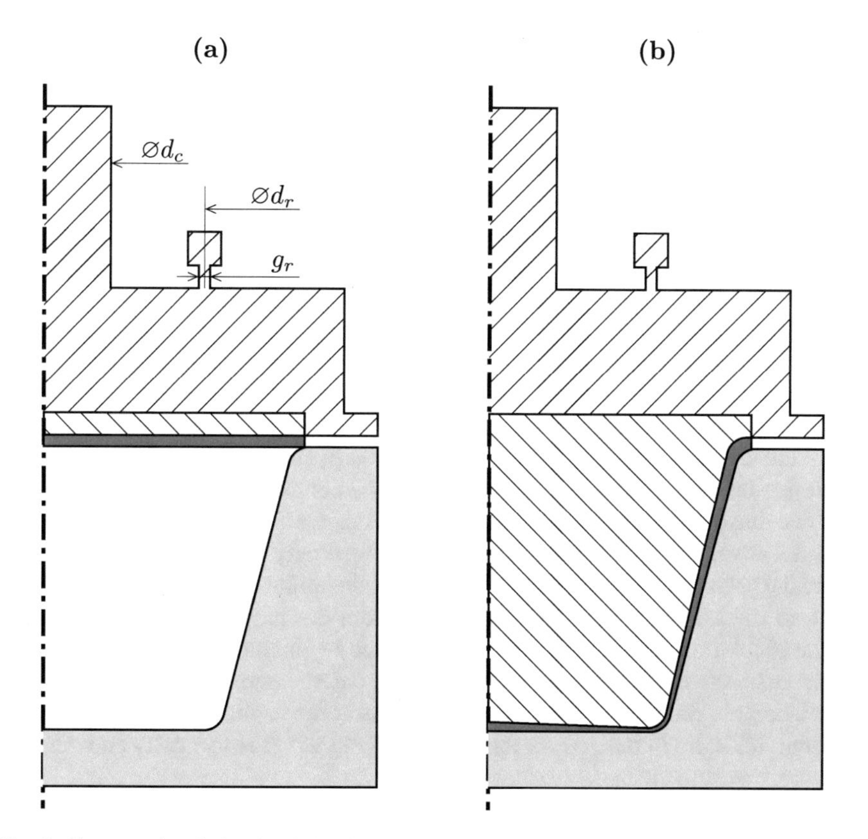

Fig. 4 Cross sectional sketch of the simulated domain (**a**) at process start and (**b**) at process end. The fluid parts are hatched, whereas the structural parts are filled in light (mold) and dark gray (plastic sheet). For better visibility of the indicated qualitative wall thickness distribution, the plastic sheet thickness is shown disproportionately high

main flow regions. As the remeshing algorithm only works for tetrahedral meshes, a mesh of said type is mandatory for the deforming zone. Initially, the whole mesh consists of approximately 10^6 cells. There are two inlets into the fluid domain: a central nozzle with the diameter d_c, and a circumferential ring nozzle with the ring diameter d_r and the gap width g_r as shown in Fig. 4a. The air in the pressure-box is modeled as an ideal gas, using further assumptions of the kinetic theory of gases, and with a three-parameter Sutherland formula for viscosity correction. For the given modeling approach, the maximum errors in comparison to the literature data [8] for dry air in the considered temperature range are less than 2% for the specific heat capacity, density and dynamic viscosity, and at maximum around 7% for the thermal conductivity. Pressure-velocity coupling is achieved by the PISO algorithm for compressible flows with segregated solutions for the flow variables. The time step size of the transient simulation is necessarily adopted from the coupling service as 10^{-4} s and therefore cannot be set explicitly. Second order discretization has been used for discretization in space and in time, but if remeshing occurs, time accuracy can fall back to first order [1]. As in the structural part, gravitational influence is neglected.

Initial and Boundary Conditions For the first simulations, the central nozzle was neglected and the corresponding boundary is treated as a wall. Inflow occurred only through the ring nozzle, so that the initially most critical part of the mesh, at the ring nozzle's gap, could be investigated regarding instabilities. Until detailed measurement results are available, that allow for treatment as a pressure boundary, the ring nozzle inlet is treated as a mass flow inlet. This is due to the fact that by imposing a mass flow inlet condition, the pressure buildup in the fluid domain causes the volumetric inflow rate to decrease gradually, which is also expected to be the qualitative behavior in the real process. Using Fluent's capability to include user-written code, the mass flow is increased from zero to the target value during the first time steps in order to avoid oscillations, that otherwise would occur due to the sudden change in inlet velocity.

As the simulated process is inherently unsteady, no stationary state simulation result can be chosen as initial condition, but the conditions in the thermoforming machine have to be estimated. The initial temperature distribution inside the pressure chamber is prescribed to decay exponentially from the TFSI interface, where the prescribed temperature is the same as the initial temperature of the plastic sheet, to the ambient temperature far away from the plastic sheet. This helps to reduce oscillations in the temperature field at the beginning of the simulation, that would arise due to the enormous temperature gradient across the TFSI interface if the whole fluid domain had ambient temperature at the beginning. The velocity and pressure fields are initialized for the whole domain with zero velocity and ambient pressure, respectively.

Turbulence Modeling The SST-k-ω turbulence model is used for turbulence modeling. It is a two-equation turbulence model, based on the eddy viscosity hypothesis for the calculation of the Reynolds stresses that provides a reasonable compromise between accuracy and computational requirements for simulations of heat transfer of impinging jets [10].

In addition to the equations for velocity and pressure, two transport equations are solved for the turbulent frequency ω and the (isotropic) turbulent kinetic energy k. SST-k-ω is an enhancement to the BSL-k-ω model, which itself is a blend between the well-known k-ε model for wall-distant flow and k-ω-model for near-wall flow. Other than the BSL-k-ω model, the SST-k-ω model contains a limiter in the formula for the turbulent viscosity with the aim to account for proper transport of turbulent shear stresses.

At stagnation points, most standard turbulence models tend to overestimate the turbulent kinetic energy, which in turn causes excessive heat transfer rates. For flows with stagnation points, like impinging jets, Fluent therefore offers the possibility to include a further limiter of the turbulent kinetic energy production, which is switched on for the TFSI model. Furthermore, the production term is enhanced in order to account for streamline curvature.

More general turbulence modeling approaches like Large Eddy Simulation allow for substantially higher accuracy in impinging jet flow simulations [10], but the small time step sizes that these methods require would render simulations of this kind infeasible, considering that not only the fluid but also the structural part would have to be solved for significantly smaller time steps. Especially with regard to future parameter studies, which will require simulation results from many different configurations as data basis, a simpler and therefore less accurate turbulence modeling approach has to be used.

3.2 *Structure Simulation*

For the structural simulation, namely the deformation of the plastic sheet into the mold, ANSYS Mechanical is used. The driving force of the deformation are the pressure forces exerted on the upper side of the plastic sheet by the fluid domain, but convective cooling also plays an important role. Other than depicted in Fig. 4, the mold is only modeled as a rigid surface, and not as a solid body. Rigidity can be assumed due to the substantially higher stiffness of the mold material compared to the plastic. Heat transfer across the contact is not used in the model at its current state, but it could also be included using a surface-only representation of the mold by a prescribed heat transfer coefficient and a prescribed temperature of the mold walls.

In some previous thermoforming simulations, instant sticking of the plastic to the mold wall at contact has been assumed (e.g. [9]). Although this appears not to be exactly the case and also other models, e.g. using Coulomb friction law have been used [4], preparatory simulations led to the conclusion that the error by assuming instant wall-sticking is negligible. In the current model, this effect is obtained by using a frictional contact with a very high friction coefficient of 1.5. This cannot completely prevent sliding, but limits it to negligible distances. An exception is at the upper corner blend radius, where there is noteworthy penetration and sliding due to the small normal stiffness imposed in order to reduce the computational cost.

The particular difficulty of the structural part is that it has to cope with three types of nonlinearity: geometrical nonlinearities arise from the high deformation, a nonlinear material constitutive equation is used and there is a large contact area. As contact pairs are evaluated on one processor and not further distributed, this heavily impairs the speed of solution of the structural part. In order to improve the parallel efficiency, therefore the model was divided along the circumference into four wedges with a corner radius of 90° each. Partitioning into more than four wedges would have been possible, but the execution speed did not improve further, in contrast the mesh quality deteriorated due to the more acute angle at the wedges' inner corner. A division along the radial direction into concentric disks was not possible, because for contact definition the surfaces for which contact is to be considered must be known before simulation start. This would require knowledge of the exact position where which part of the plastic sheet touches the mold wall. But this information is not available before the simulation.

While previous thermoforming simulations usually use shell-type elements for the discretization of the plastic sheet, in ANSYS using thermal and dynamic coupling at the interface is only supported with continuum elements (namely the 20-node hexahedral and 10-node tetrahedral SOLID226 and SOLID227 elements). Although the documentation does not give a reason for this restriction, the authors assume that this is due to the linear temperature profile that the possible shell elements assume, which may compromise accuracy in calculations that require the temperature gradient, e.g. for heat flow calculations at the TFSI interface. Said restriction has the big disadvantage that continuum elements with huge aspect ratios have to be used, otherwise the node count would reach levels prohibitively for simulations.

The large contact area limits scalability and the inevitable use of continuum elements for the thin plastic sheet causes a high computing demand, which in combination results in the structural part determining the simulation speed.

Material Modeling For the material characterization, a uniaxial tensile test is used. Some results are shown in Fig. 5. While the uniaxial tests are conducted, a novel biaxial test bench is put into service at Fraunhofer IVV Dresden, with the unique feature that it will allow for completely homogeneous biaxial straining. These biaxial tensile tests will later be used for material characterization, as these are supposed to reproduce a straining behavior which is closer to the one encountered in the thermoforming process, at least in the bottom wall of the end product [6]. For early versions, a third order Mooney-Rivlin material with Prony relaxation was fitted to the tensile test data. This turned out not be suitable any more for the complex material behavior (e.g. necking), that especially some semi-crystalline plastics exhibit and which is very important to be captured for the thermoforming process. Therefore, the material model has been switched to MISO, which is in effect a stress-strain table for given temperatures, with assumed linear elastic behavior up to the first data point and plastic straining for further data points. Between the data tables for different temperatures, interpolation is applied. The stress-strain data pairs for MISO are extracted from the measured data for a given temperature.

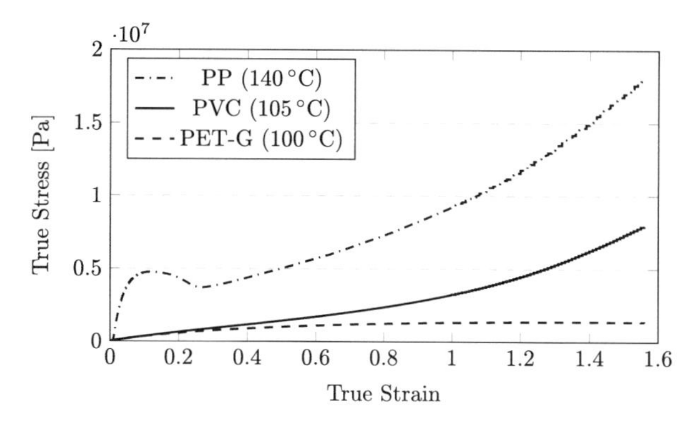

Fig. 5 Measurement results of true stress over true strain for polypropylene (PP), polyvinyl chloride (PVC), and polyethylene terephthalate (PET-G)

Initial and Boundary Conditions Due to gravity being neglected, no sagging occurs, therefore the sheet initially has the form of a plain disk. The temperature is set to a material-dependent temperature of high deformability uniformly for the whole sheet. At the outer boundary, the node displacement is fixed to zero. This implies that no further material can flow into the mold after the start, but is consistent with the wall-sticking assumption for the whole structural model. The upper side of the disk-shaped plastic sheet is the TFSI interface, the lower side is constrained only by contact with the mold surface. Air which is initially between the mold and the plastic sheet escapes during the process through small venting holes in the bottom of the mold. Their small diameters cause friction losses, which in consequence lead to pressure build-up inside the mold. Compared with the overpressure above the plastic sheet from the fluid domain, this pressure is deemed to only have very little effect and therefore is not incorporated into the model.

3.3 Coupling of Fluid and Structural Part

Coupling is done via the ANSYS System Coupling module. As depicted in Fig. 6, at the interface happens the data exchange between fluid and structural part. Transferring the near wall temperature and the heat transfer coefficient instead of the heat flow is usually more stable, but proved to be slower to converge. The model's configuration yielded convergence in three to four coupling steps per time step.

Convergence of the coupling is determined using a global measure: the root mean square of the normalized change in transferred values, computed for a quantity ϕ according to Eq. (1), with n being the current and $n-1$ being the previous coupling

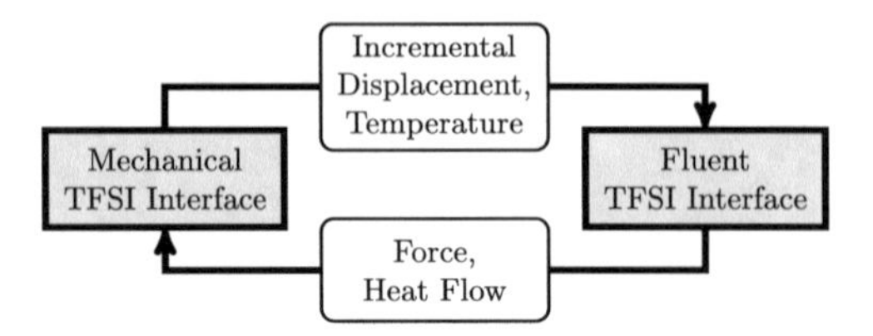

Fig. 6 Transferred quantities at the TFSI interface

iteration counter. The overbar indicates averaging with the arithmetic mean.

$$RMS_{normalized} = \sqrt{\overline{\left(\frac{\phi^n - \phi^{n-1}}{0.5(max(abs(\phi^n) - min(abs(\phi^n)) + \overline{abs(\phi^n)})}\right)^2}} \tag{1}$$

If this relative change to drop below 1%, the coupling loop for the given time step is considered converged and the simulation proceeds with the next time step.

The time step has been chosen according to prior fluid simulations, in which the SST-k-ω-model and a scale-adaptive variant have been tested for stability. While SST-k-ω results did not depend on the time step size for values below around 10^{-4} s, scale-adaptive simulations still showed dependence on the time step size and would have required at maximum half the value of SST-k-ω, which appeared to be too restrictive, having in mind that each time step consists of several coupling iterations. Fluid and structural part of the simulation automatically adopt the time step size of 10^{-4} s, which remains constant for the whole simulation.

At the beginning of the simulation, the mapping between fluid and structural part is established once and then used for the whole simulation, therefore a remeshing of the interface surface mesh is not possible. The fluid mesh is notably finer, therefore a non-conformal mesh interface is used. Non-conserved quantities like incremental displacement and temperature are transferred using a profile-preserving approach. For conserved quantities like forces and heat flow, a profile-preserving algorithm that also guarantees conservation is used.

The simulations have been using 20 processor cores for the fluid part and 16 processor cores for the structural part on the *emmy* cluster at Regionales RechenZentrum Erlangen (RRZE). Although the fluid part scales well, only an insignificant speed gain can be obtained from using more processor cores, because most of the time is spent in the structural computations with very limited scalability.

4 Simulation Results

In Fig. 7, some simulation results at different time steps are shown. The figure has been created by mapping all nodes around the circumference onto one plane and connecting them by increasing radius by lines, without taking into account the actual elemental connectivity. This allows for a good approximation and averaging where

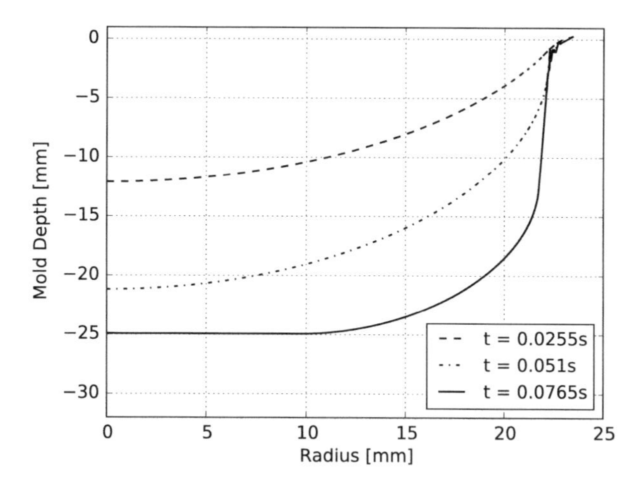

Fig. 7 Deformation at different time steps

there are no big deviations across the circumference, but at the downside causes artifacts to appear where this does not apply. One can clearly see in Fig. 7 that this is the case at the upper radius, where highest penetration and inaccuracies in the contact calculations due to the rather low contact normal stiffness occur.

As one can see in Fig. 7, before the plastic sheet gets in contact with the mold's bottom wall, it touches the side walls starting from the upper blend radius. In this region, the wall thickness has usually rather high values, because at this time and location the stretching rate is only moderate. At the last time step shown, already a large area is in contact with the mold walls, and according to the wall-sticking assumption, will undergo virtually no further stretching. This explains why at the lower corner blend radius the lowest wall thickness is to be expected.

For the first time step shown in Fig. 7, a snapshot of the temperature profile is visualized with a contour plot in Fig. 8. The notable difference between the mean air temperature in the pressure-box and the inflowing air which is at ambient temperature is partially due to compression, but mainly due to the initial condition of the temperature field as explained in Sect. 3.1. One can see how the air jet leaving the ring nozzle leads to a high temperature gradient (visible as compressed temperature contours) on the plastic sheet surface in the impingement region. There, the jet is deflected, so that a wall-parallel flow region develops. Close to the symmetry axis, the wall-parallel flow is redirected upwards again, leaving a region of low temperature gradient (visible as extended temperature contours) at the middle of the plastic sheet. The described behavior is most pronounced at process start, when the jet traveling distance is smallest. During the deformation process, as the distance between the nozzle and the plastic sheet increases, the cooling influence of the jet diminishes.

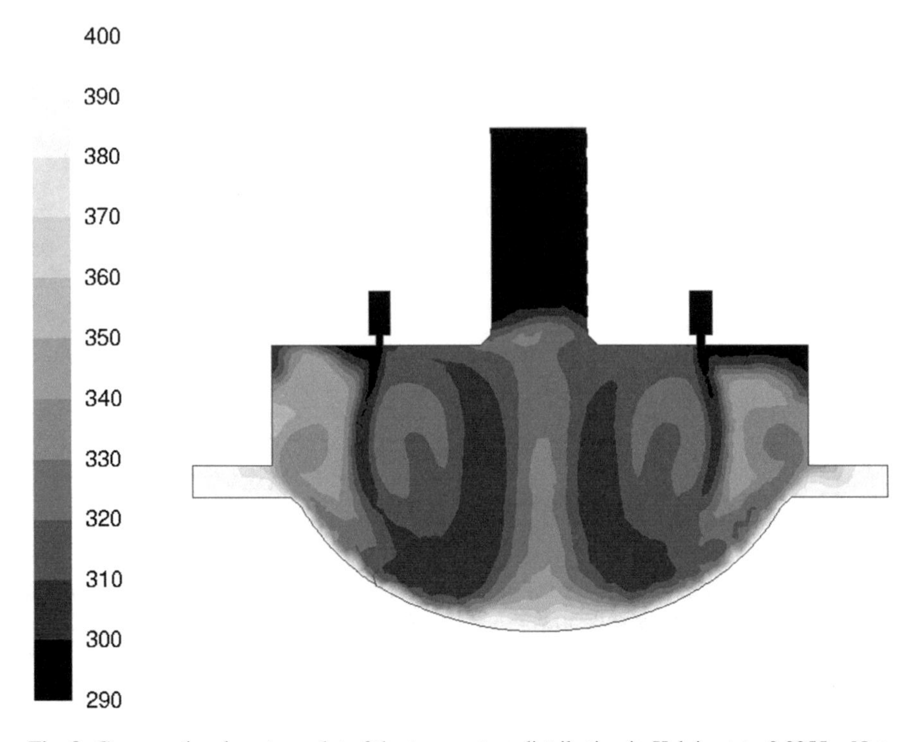

Fig. 8 Cross sectional contour plot of the temperature distribution in Kelvin at t = 0.0255 s. Note that inflow only occurs through the ring nozzle, the central nozzle was treated as an impermeable wall

The simulated process results so far are not yet representative for the parameter studies that will be conducted after the model has been calibrated and validated. This explains why the computed wall thickness distributions don't differ significantly from the wall thickness distributions of the common process.

So far the simulations were done using a sheet thickness of 0.25 mm, which is about half as thick as the simulations for the validation will be. This means that for the upcoming simulations with increased thickness, more convenient conditions prevail than for the preliminary test simulations. This could alleviate the most severe challenges, especially the high aspect ratio of the continuum elements.

5 Experimental Results

Parallel to the setup of the computational model, experiments have been conducted, so that experimental data for the axis-symmetric test mold is already available. The first results underpin the authors' presumption that the wall thickness distribution can be positively influenced by directed jets of pressurized air with the geometrical configuration, namely using a central nozzle and a ring nozzle. For these experiments, at process start only one of the two nozzles opened, and the other one was switched on during the process.

Contrary to the authors' initial hypothesis that cooling at a certain radius with the ring nozzle would lead to the most notable improvements, the best results were obtained by starting with inflow through the central nozzle only. For PET-A material, the minimum measured wall thickness, located at the lower blend radius, has improved by around 40% compared to the common process. This corresponds to an absolute improvement in wall thickness of 0.02 mm.

6 Conclusion and Outlook

A TFSI model of the thermoforming process has been set up using the ANSYS software suite. It includes dynamic as well as thermal coupling between the pressurized-air flow in the pressure-box and the deforming plastic sheet.

Further effort will be made in order to complete the coupled simulation model so that the whole thermoforming process can be simulated. Special focus in improving the fluid part will be on remeshing quality and calibration of the wall heat transfer modeling. For the structural part, limitations by the software can only be alleviated by relaxing the accuracy requirements. The main problems are that no shell elements are available for TFSI and that the contact area is large compared to the overall structural model size, which in combination leads to the structural part determining the simulation speed. Reductions in computing time and an increase in stability of the structural part are expected for the upcoming validation simulations. This is due to the higher sheet thicknesses used in the experiments and therefore lower aspect ratios of the used continuum elements than in previous simulation runs. Upon completion and validation of the computational model, parameter variations are planned for the inlet mass flow rates of both nozzles, for the mold geometry and for the nozzle geometries. Backup through experimental investigations is indispensable, because the high number of assumptions in the model, necessary in order to maintain an applicable level of complexity require careful validation.

While completion and validation of the model are still pending, experimental results with simply shaped nozzles have confirmed the potential of using directed flow of the pressurized compressed-air for significant improvements in the homogeneity of the wall thickness distribution. Further experimental parameter studies will be conducted for variations in the impingement angle and geometry variations. Future experiments and simulations using a second, non-axis-symmetric mold shape will test for the applicability to more complex geometries.

The authors are convinced that with both a representative computational model and a sound experimental data base, realistic estimations of the impact of the different parameters on the wall thickness distribution can be derived, which can then be used for process optimization.

Acknowledgements The IGF project 18536 BG of the Industry Association for Food Technology and Packaging (IVLV) was funded via AiF within the funding program of Collective Industrial Research (IGF) from the German Federal Ministry for Economic Affairs and Energy (BMWi) on the basis of a resolution of the German Bundestag.

References

1. ANSYS (R) Academic Research CFD, Release 17.2, Help System, Fluent User's Guide 10.6.2.5. ANSYS, Inc
2. Carlone, P., Palazzo, G.S.: Finite element analysis of the thermoforming manufacturing process using the hyperelastic mooney-rivlin model. In: Gavrilova, M., et al. (eds.) International Conference on Computational Science and its Applications (ICCSA 2006). Lecture Notes in Computer Science, vol. 3980, pp. 794–803. Springer, Berlin (2006)
3. Choo, H.L., Martin, P.J., Harkin-Jones, E.M.A.: Measurement of heat transfer for thermoforming simulations. Int. J. Mater. Form. **1**(Suppl. 1), 1027–1030 (2008). https://doi.org/10.1007/s12289-008-0233-7
4. Dong, Y., Lin, R.J.T., Bhattacharyya, D.: Finite element simulation on thermoforming acrylic sheets using dynamic explicit method. Polym. Polym. Compos. **14**(3), 307–328 (2006)
5. Kayatz, F., Claus, R.: Strömungsoptimiertes Thermoformen—Einfluss von Düsengeometrie und Formdruck auf die Wanddickenverteilung, Verpackungs-Rundschau, pp. 70–71 (2013)
6. McCool, R., Martin, P.J.: Thermoforming process simulation for the manufacture of deep-draw plastic food packaging. Proc. Inst. Mech. Eng. E J. Process Mech. Eng. **225**, 269–279 (2011)
7. Min, S.C., Zhang, H.Q., Yang, H.-J.: Thermoformed container wall thickness effects on orange juice quality. J. Food Process. Preserv. **35**, 758–766 (2011). https://doi.org/10.1111/j.1745-4549.2011.00526.x
8. VDI e.V. (VDI-GVC): VDI-Wärmeatlas, 11. Auflage. Springer, Berlin (2013). https://doi.org/10.1007/978-3-642-19981-3
9. Warby, M.K., Whiteman, J.R., Jiang, W.-G., Warwick, P., Wright, T.: FE simulation of thermoforming processes for polymer sheets. Math. Comput. Simul. **61**, 209–218 (2003)
10. Zuckerman, N., Lior, N.: Impingement heat transfer: correlations and numerical modeling. J. Heat Transf. **127**, 544 (2005). https://doi.org/10.1115/1.1861921

Monolithic Simulation of Convection-Coupled Phase-Change: Verification and Reproducibility

Alexander G. Zimmerman and Julia Kowalski

Abstract Phase interfaces in melting and solidification processes are strongly affected by the presence of convection in the liquid. One way of modeling their transient evolution is to couple an incompressible flow model to an energy balance in enthalpy formulation. Two strong nonlinearities arise, which account for the viscosity variation between phases and the latent heat of fusion at the phase interface.

The resulting coupled system of PDE's can be solved by a single-domain semi-phase-field, variable viscosity, finite element method with monolithic system coupling and global Newton linearization (Danaila et al., J Comput Phys 274:826–840, 2014). A robust computational model for realistic phase-change regimes furthermore requires a flexible implementation based on sophisticated mesh adaptivity. In this article, we present first steps towards implementing such a computational model into a simulation tool which we call Phaseflow (Zimmerman, https://github.com/geo-fluid-dynamics/phaseflow-fenics).

Phaseflow utilizes the finite element software FEniCS (Alnæs et al., Arch. Numer Softw 3(100):9–23, 2015), which includes a dual-weighted residual method for goal-oriented adaptive mesh refinement. Phaseflow is an open-source, dimension-independent implementation that, upon an appropriate parameter choice, reduces to classical benchmark situations including the lid-driven cavity and the Stefan problem. We present and discuss numerical results for these, an octadecane PCM convection-coupled melting benchmark, and a preliminary 3D convection-coupled melting example, demonstrating the flexible implementation. Though being preliminary, the latter is, to our knowledge, the first published 3D result for this method. In our work, we especially emphasize reproducibility and provide an easy-to-use portable software container using Docker (Boettiger, ACM SIGOPS Oper Syst Rev 49(1):71–79, 2015).

A. G. Zimmerman (✉) · J. Kowalski
AICES Graduate School, RWTH Aachen University, Aachen, Germany
e-mail: zimmerman@aices.rwth-aachen.de

M. Schäfer et al. (eds.), *Recent Advances in Computational Engineering*, Lecture Notes in Computational Science and Engineering 124,
https://doi.org/10.1007/978-3-319-93891-2_11

Keywords Incompressible flow · Finite element method · Newton method · Phase-change · Reproducibility

Nomenclature

t	Time
t_f	Final time
$\mathbf{x}$	Spatial coordinates $\mathbf{x} = \begin{pmatrix} x & y & z \end{pmatrix}$
x^*	Phase-change interface's position
p	Pressure field
$\mathbf{u}$	Velocity field
T	Temperature field
$\mathbf{f}_B$	Buoyancy force
r	Regularization smoothing factor
T_r	Regularization central temperature
Pr	Prandtl number
Ra	Rayleigh number
Ste	Stefan number
μ	Dynamic viscosity
ϕ	Semi-phase-field
Δt	Time step size
$()_n$	Values from discrete time n
γ	Coefficient for penalty stabilization
h	Finite element cell diameter
$\mathbf{F}$	The vector-valued strong form
F	Functional for the variational form
Ω	The spatial domain
$\partial\Omega$	Boundary of the spatial domain
$\partial\Omega_T$	Dirichlet boundary for T
V	The scalar solution function space
$\mathbf{V}$	The vector solution function space
$\mathbf{W}$	Mixed finite element function space
$\mathbf{w}$	System solution $\mathbf{w} = \begin{pmatrix} p & \mathbf{u} & T \end{pmatrix}$
$\boldsymbol{\psi}$	Finite element basis functions
M	Adaptive goal functional
ϵ_M	Adaptive solver tolerance
$\delta\mathbf{w}$	Residual of linearized system
ω	Newton method relaxation factor
$()^k$	Values from Newton iteration k

1 Introduction

The melting and solidification of so-called phase-change materials (PCM's) are relevant to many applications ranging from the design of latent heat based energy storage devices [11], to ice-ocean coupling and its effects on Earth's climate [9], to the evolution of ocean worlds on the icy moons of our solar system [13] and the design of robotic melting probes for their exploration [15]. The predictive modeling of phase-change systems is, however, challenging due to (1) strong nonlinearities at the phase-change interface (PCI), (2) the coupling of several physical processes (i.e multi-physics), and (3) a large range of relevant scales both in space and time (i.e. multi-scale). Any mathematical model of a complex phase-change process hence manifests as a multi-scale and multi-parameter, nonlinear PDE system. We aim to develop a robust and flexible model to simulate these systems.

In this work, we will focus on phase-change in the presence of liquid convection. Convection can have a tremendous effect on the evolution of phase-interfaces, as shown in Fig. 1. A comprehensive introduction to melting and freezing without convection is given in [1]. A mathematical model that accounts for convection-coupled phase-change is presented in [4]. Therein mushy layer theory is introduced, which provides a model for understanding the PCI at macro-scale. The physical system is mathematically modeled by considering balance laws for mass, momentum, and energy, with the momentum and energy balances coupled via buoyancy (which forces natural convection).

This constitutes a difficult multi-physics problem. In a PCM domain, certain physics dominate in the solid and liquid subdomains. To solve the coupled problem, out of many approaches in the literature, we employ the single-domain semi-phase-field enthalpy method. With this approach, the convection itself is handled by an incompressible Navier-Stokes flow model, while the energy balance is modeled as the convection and diffusion of an enthalpy field. The key idea is to solve the same equations on the entire domain. This is commonly referred to as the *fixed grid*

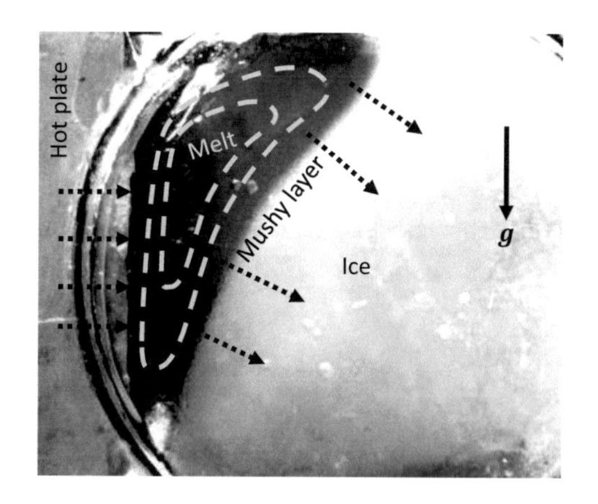

Fig. 1 Sketch of the physical model superimposed on a melting experiment [16]. Heating from the left advances the phase-change interface (PCI) rightward. Buoyancy forces a convection cell, causing the upper PCI to advance more rapidly. In the absence of convection, the PCI would propagate as a planar front, indicating the significant impact of convection on the PCI's evolution

approach, for which various techniques are reviewed in [18]. Using an enthalpy method, phase-change latent heat effects are isolated to a source term in the energy balance [17]. Single-domain methods require a velocity correction scheme, for which we select the variable viscosity method, thereby treating the solid as a highly viscous fluid.

Efficiently applying this approach requires local mesh refinement to resolve the moving PCI. This can be accomplished either with mesh refinement or front-tracking [11]. Adopting the former approach, we leverage the existing work on goal-oriented adaptive mesh refinement (AMR) methods, particularly the dual-weighted residual method [3]. These days, goal-oriented AMR is widely practiced, and multiple open-source software libraries provide this capability.

In the following, Sect. 2 describes the mathematical model. Section 3 presents numerical methods for the model's discretization and linearization. Section 4 presents our implementation, Phaseflow [20], based on the finite element software FEniCS [2]. Finally, Sect. 5 presents verification via comparison to benchmark problems, a convergence study for the 1D Stefan problem, and a preliminary result for convection-coupled melting in a 3D domain. We close with conclusions and an outlook.

2 Mathematical Phase-Change Model

2.1 The Governing Equations

For the coupled phase-change system, out of many approaches from the literature, we adopt an enthalpy formulated [17], single-domain semi-phase-field [5], variable viscosity model. The mass and momentum balances (1) and (2) are given by the incompressible Navier-Stokes equations, with velocity field $\mathbf{u} = \mathbf{u}(\mathbf{x}, t)$ and pressure field $p = p(\mathbf{x}, t)$. Invoking the Boussinesq approximation, we extend the momentum equation with a temperature-dependent buoyancy forcing term $\mathbf{f}_B(T)$ which couples it to the energy equation. This approach is well established in the context of natural convection [19]. Furthermore, we consider the phase-dependent viscosity $\mu(\phi)$ in the momentum equation, where $\phi = \phi(T)$ is the temperature-dependent phase. For constant heat capacity, the energy balance in enthalpy form reduces to (3), which is an extended form of the convection-diffusion equation for the temperature field $T = T(\mathbf{x}, t)$. The diffusion term in (3) is scaled by the Prandtl number Pr. The nonlinear source term $\frac{1}{\text{Ste}}\frac{\partial}{\partial t}\phi$ accounts for the phase-change latent heat, where Ste is the Stefan number. Altogether the system of governing equations is

$$\nabla \cdot \mathbf{u} = 0 \tag{1}$$

$$\frac{\partial}{\partial t}\mathbf{u} + (\mathbf{u} \cdot \nabla)\,\mathbf{u} + \nabla p - \nabla \cdot (2\mu(\phi)\mathbf{D}(\mathbf{u})) + \mathbf{f}_B(T) = 0 \tag{2}$$

$$\frac{\partial}{\partial t}T - \frac{1}{\text{Ste}}\frac{\partial}{\partial t}\phi + \nabla \cdot (T\mathbf{u}) - \frac{1}{\text{Pr}}\Delta T = 0 \tag{3}$$

where the symmetric part of the rate-of-strain tensor is $\mathbf{D}(\mathbf{u}) = \frac{1}{2}\left(\nabla\mathbf{u} + (\nabla\mathbf{u})^{\mathrm{T}}\right)$. The equations are unitless per the normalization in [8], shifted always such that $T = 0$ corresponds to the temperature of fusion.

We refer to ϕ as a *semi*-phase-field [5], because we do not treat ϕ as an additional unknown. Rather, ϕ maps the temperature field to values between zero and one. For this we use

$$\phi(T) = \frac{1}{2}\left(1 + \tanh\frac{T_r - T}{r}\right), \tag{4}$$

where T_r is the central temperature and r is a smoothing parameter. The Newton method requires the differentiability of (4). Subject to this requirement, there are many other useful regularizations from which to choose. Figure 2 plots ϕ and ϕ'. For the Stefan problem in Sect. 5.3, we simply set T_r equal to the physical temperature

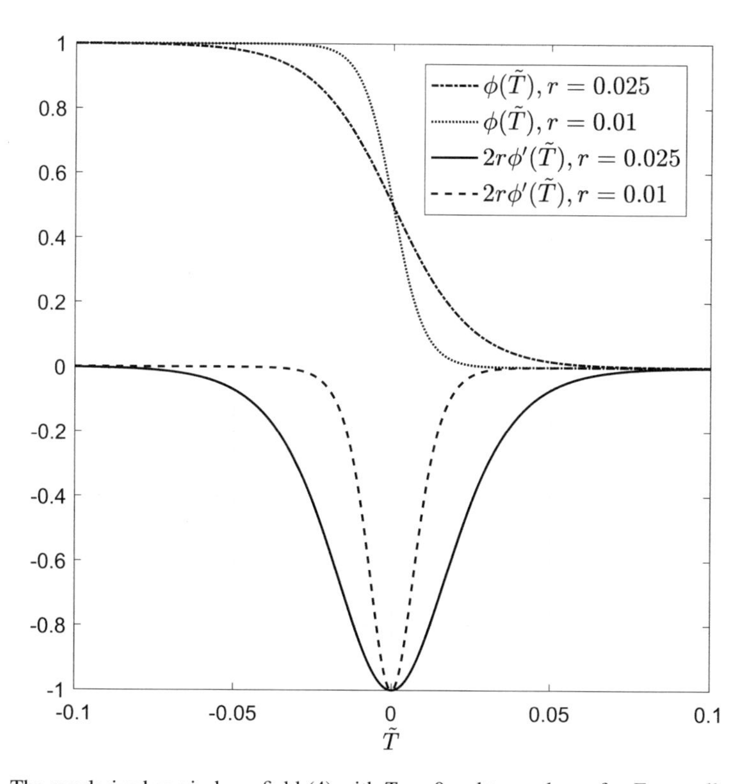

Fig. 2 The regularized semi-phase-field (4) with $T_r = 0$ and two values of r. For smaller r, $\phi(T)$ steepens and its derivative $\phi'(T)$ approaches the Dirac delta function. Here the derivative is scaled by $2r$ for convenience

of fusion (i.e. $T_r = 0$ for the normalized system). Following [8], for the coupled convection problem, we set T_r such that the strongest variation is localized near the newly appearing phase, e.g. liquid for melting. Physically, the region where $0 < \phi < 1$ is analogous to a mushy layer. In this sense, we can view ϕ as the solid volume-fraction. Given ϕ, we define the phase-dependent viscosity as functions of the liquid and solid values (assuming constants μ_L and μ_S respectively) with

$$\mu(\phi) = \mu_L + (\mu_S - \mu_L)\phi, \tag{5}$$

With the variable viscosity method, the solid is treated as a highly viscous fluid, forcing its flow velocity to zero. In general, not only viscosity varies in the PCM domain. For example, the thermal material properties of water-ice vary significantly between the solid and liquid phases. In the case of octadecane PCM's, such as the benchmark in Sect. 5.4, we may treat these material properties as constants [19].

2.2 *The Initial Boundary Value Problem*

To simulate the time evolution of the PCM system, we will solve (8) as an initial boundary value problem subject to the initial values

$$\left(p(\mathbf{x}, t), \mathbf{u}(\mathbf{x}, t), T(\mathbf{x}, t)\right)^T = \left(0, \mathbf{u}_0(\mathbf{x}), T_0(\mathbf{x})\right)^T \quad \forall(\mathbf{x}, t) \in \Omega \times 0 \tag{6}$$

and boundary conditions

$$\begin{aligned} \mathbf{u}(\mathbf{x}, t) &= \mathbf{u}_D(\mathbf{x}, t) \quad \forall(\mathbf{x}, t) \in \partial\Omega \times (0, t_f], \\ T(\mathbf{x}, t) &= T_D(\mathbf{x}, t) \quad \forall(\mathbf{x}, t) \in \partial\Omega_T \times (0, t_f], \\ (\hat{\mathbf{n}} \cdot \nabla)T(\mathbf{x}, t) &= 0 \quad \forall(\mathbf{x}, t) \in (\partial\Omega \setminus \partial\Omega_T) \times (0, t_f] \end{aligned} \tag{7}$$

For the applications in Sect. 5, we apply for all boundaries the velocity no-slip condition. This implies homogeneous Dirichlet boundary conditions, $\mathbf{u}_D = \mathbf{0}$, for all cases except the lid-driven cavity in Sect. 5.1, where the "moving" lid makes $\mathbf{u}_D$ non-homogeneous. The notation $\partial\Omega_T$ refers to the boundary subdomain with non-homogeneous Dirichlet boundary conditions on the temperature. Physically this models a thermal reservoir beyond the wall, keeping the wall at a constant temperature. All other walls are prescribed homogeneous Neumann boundary conditions on the temperature, which physically models thermally insulated (i.e. adiabatic) walls.

The generality of this initial boundary value problem allows us to solve a variety of interesting benchmarks by setting the appropriate parameters. In Sect. 5 we will demonstrate the 2D lid-driven cavity, 2D heat-driven cavity, 1D Stefan problem, 2D convection-coupled melting of an octadecane PCM, and 3D convection-coupled melting. Our implementation which we present in Sect. 4 is as versatile as the model.

Testing these benchmarks is accomplished with short Pythons scripts specifying the parameters.

3 Numerical Methods

We base our work on the numerical approach from [8]. Therefore, we discretize in time via finite differences, discretize in space via the finite element method (FEM) with a penalty formulation for stabilization, couple the system monolithically, and solve the nonlinear system globally via Newton's method.

3.1 The Nonlinear Variational Problem

We denote the system's solution as $\mathbf{w} = \begin{pmatrix} p & \mathbf{u} & T \end{pmatrix}$ and write the strong form (1), (2), and (3) as the vector-valued functional

$$\mathbf{F}(\mathbf{w}) = \mathbf{0} \quad \forall (\mathbf{x}, t) \in \Omega \times (0, t_f], \tag{8}$$

with the three components respectively being the mass, momentum, and energy equations. To stabilize the finite element method, following the penalty formulation in [8], we add a pressure stabilizing term γp to the mass component. The coefficient γ is generally small, and in this case is $\gamma = 10^{-7}$. We consider 1D, 2D, and 3D spatial domains, respectively allowing for $\Omega \subset \mathbb{R}^1$, $\Omega \subset \mathbb{R}^2$, or $\Omega \subset \mathbb{R}^3$.

Following the standard Galerkin finite element method, we write the variational (i.e. weak) problem whose solution approximates (8). To obtain the monolithic system, we use mixed finite elements [7]. Therefore, we multiply (8) from the left by test functions $\boldsymbol{\psi} = \begin{pmatrix} \psi_p & \boldsymbol{\psi}_u & \psi_T \end{pmatrix}$ and integrate by parts over the domain. Finally, we employ the fully implicit Euler method by substituting the time discretizations $\partial_t \mathbf{u} = \Delta t^{-1}(\mathbf{u}_{n+1} - \mathbf{u}_n)$, $\partial_t T = \Delta t^{-1}(T_{n+1} - T_n)$, $\partial_t \phi = \Delta t^{-1}(\phi_{n+1} - \phi_n)$.

Altogether, for homogeneous Neumann boundary conditions, this yields the time-discrete nonlinear variational form

$$\begin{aligned} F(\boldsymbol{\psi}, \mathbf{w}) = \int_\Omega \begin{pmatrix} \psi_p & \boldsymbol{\psi}_u & \psi_T \end{pmatrix} \mathbf{F}(p, \mathbf{u}, T) d\mathbf{x} = \\ b(\mathbf{u}, \psi_p) - (\psi_p, \gamma p) \\ + \left(\boldsymbol{\psi}_u, \frac{1}{\Delta t}(\mathbf{u}_{n+1} - \mathbf{u}_n) + \mathbf{f}_B(T) \right) + c(\mathbf{u}; \mathbf{u}, \boldsymbol{\psi}_u) + b(\boldsymbol{\psi}_u, p) + a(\mu(\phi); \mathbf{u}, \boldsymbol{\psi}_u) \\ + \frac{1}{\Delta t} \left(\psi_T, T_{n+1} - T_n - \frac{1}{\text{Ste}} (\phi(T_{n+1}) - \phi(T_n)) \right) + \left(\nabla \psi_T, \frac{1}{\text{Pr}} \nabla T - T\mathbf{u} \right) \end{aligned} \tag{9}$$

where we use the short-hand $(u, v) = \int_\Omega uv d\mathbf{x}$, $(\mathbf{u}, \mathbf{v}) = \int_\Omega \mathbf{u} \cdot \mathbf{v} d\mathbf{x}$ for integrating inner products. Additionally, we use a common notation [10] for the linear, bilinear, and trilinear forms of the variational Navier-Stokes equations

$$
\begin{aligned}
a &: \mathbf{V} \times \mathbf{V} \to \mathbb{R}, \quad a(\mu; \mathbf{u}, \mathbf{v}) = 2 \int_\Omega \mu \mathbf{D}(\mathbf{u}) : \mathbf{D}(\mathbf{v}) d\mathbf{x} \\
b &: \mathbf{V} \times V \to \mathbb{R}, \quad b(\mathbf{u}, p) = -\int_\Omega p \nabla \cdot \mathbf{u} d\mathbf{x} \\
c &: \mathbf{V} \times \mathbf{V} \times \mathbf{V} \to \mathbb{R}, \quad c(\mathbf{u}; \mathbf{z}, \mathbf{v}) = \int_\Omega \mathbf{v}^{\mathrm{T}} (\nabla \mathbf{z}) \mathbf{u} d\mathbf{x}
\end{aligned} \tag{10}
$$

Given (9), we write the variational problem as

Find $\mathbf{w} \in \mathbf{W}$ such that

$$F(\boldsymbol{\psi}, \mathbf{w}) = 0 \quad \forall \boldsymbol{\psi} \in \hat{\mathbf{W}} \tag{11}$$

where $\mathbf{W} = V \times \mathbf{V} \times V$ and $\hat{\mathbf{W}} = \hat{V} \times \hat{\mathbf{V}} \times \hat{V}$. This distinction comes from how non-homogeneous boundary conditions are often handled in finite element method implementations. The solution is split into homogeneous and non-homogeneous parts, the former is found, and then the latter is reconstructed. Therefore, for example, T belongs to the space $V = \{v \in H_0^1(\Omega) : v = T_D \text{ on } \partial\Omega\}$, while ψ_T belongs to the space $\hat{V} = \{v \in H_0^1(\Omega) : v = 0 \text{ on } \partial\Omega\}$, where $H_0^1(\Omega)$ is the classical Hilbert space fulfilling requirements for continuity and compact support.

The vector-valued function space $\mathbf{V}$ depends on the spatial domain's dimensionality, e.g. $\mathbf{V} = V \times V$ in 2D. For the incompressible Navier-Stokes solution, we use the Taylor-Hood element [10], i.e. we restrict the pressure solution to piece-wise continuous linear Lagrange polynomials, and we restrict the velocity solution to piece-wise continuous quadratic Lagrange polynomials. We use the same polynomial space for temperature and pressure.

3.2 Linearization via Newton's Method

We apply Newton's method by solving a sequence (indexed by superscript k) of linear problems

Find $\delta\mathbf{w} \in \mathbf{W}$ such that

$$D_\mathbf{w} F(\boldsymbol{\psi}, \mathbf{w}^k; \delta\mathbf{w}) = F(\boldsymbol{\psi}, \mathbf{w}^k) \quad \forall \boldsymbol{\psi} \in \hat{\mathbf{W}} \tag{12}$$

for the residual $\delta\mathbf{w}$ which updates the solution $\mathbf{w}^k$, converging $\mathbf{w}^k$ to the approximate solution of the nonlinear problem. The Gâteaux derivative of $F(\boldsymbol{\psi}, \mathbf{w}^k)$, defined as $D_\mathbf{w} F(\boldsymbol{\psi}, \mathbf{w}^k; \delta\mathbf{w}) \equiv \frac{d}{d\epsilon} F(\boldsymbol{\psi}, \mathbf{w}^k + \epsilon\delta\mathbf{w})|_{\epsilon=0}$, is given by

$$\begin{aligned} D_\mathbf{w} F(\boldsymbol{\psi}, \mathbf{w}^k; \delta\mathbf{w}) &= b(\delta\mathbf{u}, \psi_p) - (\psi_p, \gamma\delta p) \\ + \left(\boldsymbol{\psi}_u, \frac{1}{\Delta t}\delta\mathbf{u} + \delta T \mathbf{f}_B'(T^k)\right) &+ c(\mathbf{u}^k; \delta\mathbf{u}, \boldsymbol{\psi}_u) + c(\delta\mathbf{u}; \mathbf{u}^k, \boldsymbol{\psi}_u) \\ + b(\boldsymbol{\psi}_u, \delta p) &+ a\left(\delta T \mu'(T^k); \mathbf{u}^k, \boldsymbol{\psi}_u\right) + a\left(\mu(T^k); \delta\mathbf{u}, \boldsymbol{\psi}_u\right) \\ + \frac{1}{\Delta t}\left(\psi_T \delta T, 1 - \frac{1}{\mathrm{Ste}}\phi'(T^k)\right) &+ \left(\nabla\psi_T, \frac{1}{\mathrm{Pr}}\nabla\delta T - T^k\delta\mathbf{u} - \delta T\mathbf{u}^k\right), \end{aligned} \tag{13}$$

For a given $\mathbf{w}^k$ (13) and hence (12) are linear with respect to the unknown Newton residual $\delta\mathbf{w}$. In typical fashion, we define the test functions, solution and Newton residual as linear combinations of the same basis. Upon selecting a mesh and concrete basis, this allows (12) to be re-written as a discrete linear system of the form $\mathbf{A}\mathbf{x} = \mathbf{b}$, which can be efficiently solved on a computer by standard methods and software. In this work, we directly solve each linear system with LU decomposition (using the interface of FEniCS, discussed in Sect. 4).

We use the latest discrete time solution as the initial guess for the Newton solver, i.e. we initialize $\mathbf{w}^0 = \mathbf{w}_n$. After each iteration solving for $\delta\mathbf{w}$, we update the solution with the relaxed residual $\mathbf{w}^{k+1} := \mathbf{w}^k + \omega\delta\mathbf{w}$, where $0 < \omega \leq 1$ is a relaxation factor. Relaxing Newton's method is often useful for highly nonlinear problems. The 1D and 2D results in Sect. 5 use the full Newton ($\omega = 1$) method. So far only the preliminary 3D result requires relaxation, where we will set $\omega = 0.8$.

3.3 *Adaptive Mesh Refinement*

The single-domain approach requires local mesh refinement, or else the computational cost would quickly become impractical, especially in 3D. Furthermore, we cannot a priori prescribe where to locally refine the mesh, since our goal is to predict the position of the PCI. This means we must employ adaptive mesh refinement, now commonly referred to as AMR. The theory of AMR fundamentally requires an error estimator, which exist for many interesting problems [14]. In the context of phase-change simulations, hierarchical error estimators have been derived for the Stefan problem [5]; but no such rigorous work has been completed for the problem with coupled convection. Promising results are reported in [8], wherein a mesh adaptivity procedure by metric control which was particular to the software library and its Delaunay mesh generation procedure is used. Unfortunately, it is unclear which metrics were exactly used for adaptivity in those results.

We instead employ goal-oriented AMR [3]. To discuss AMR, let us briefly write the spatially discrete problem which is dependent on a mesh $\mathbf{W}_h \subset \mathbf{W}(\Omega)$:

Find $\mathbf{w}_h \in \mathbf{W}_h \subset \mathbf{W}(\Omega)$ such that

$$F(\boldsymbol{\psi}_h, \mathbf{w}_h) = 0 \quad \forall \boldsymbol{\psi}_h \in \hat{\mathbf{W}}_h \subset \hat{\mathbf{W}} \tag{14}$$

Goal-oriented AMR requires some goal functional $M(\mathbf{w})$ to be integrated over the domain Ω. The goal-oriented adaptive solution of (14) can then be written:

Find $\mathbf{W}_h \subset \mathbf{W}(\Omega)$ and $\mathbf{w}_h \in \mathbf{W}_h$ such that

$$|M(\mathbf{w}) - M(\mathbf{w}_h)| < \epsilon_M \tag{15}$$

where ϵ_M is some prescribed tolerance. Since $M(\mathbf{w})$ is unknown, we still require an error estimator. To this end, we use the dual-weighted residual method, as implemented in FEniCS [2]. Computing cell-wise error estimates requires solving a linearized adjoint (with respect to the goal) problem. The primal and adjoint problems are solved on a hierarchy of meshes. Computing the linearized adjoint solution on each mesh is relatively cheap compared to solving the nonlinear primal problem. See [3] for a full explanation of the method. For the adaptive solutions in this work, we set the goal functional

$$M(\mathbf{w}) = \int_{\Omega} \phi(T) d\mathbf{x} \tag{16}$$

which represents the volume of solid material remaining in the domain.

4 Implementation

The method presented in this paper, along with its application to a series of test problems, were implemented by Zimmerman into an open-source Python module named Phaseflow [20], using the finite element library FEniCS [2].

4.1 *FEniCS*

The abstract interface of FEniCS [2] makes it an ideal library to quickly prototype models and methods which use FEM. Additionally, standard algorithms for Newton linearization and goal-oriented AMR are already implemented. FEniCS is an umbrella project. A major back-end component is the C++ library DOLFIN, which implements most of the classes and methods seen by the user. Python interfaces are generated mostly automatically, and indeed FEniCS is primarily used as a Python module. From a conceptual perspective, we are most interested in two components of FEniCS: the Unified Form Lanugage (UFL) and the FEniCS Form Compiler

(FFC). UFL allows us to write down the abstract discrete variational (i.e. weak) form in a way that is understood by FEniCS. This means that to implement variational forms, we write them as source code almost character-for-character. The FEniCS Form Compiler (FFC) then automatically implements the FEM matrix assembly routine in optimized C++ code using just-in-time (JIT) compiling.

4.2 *Phaseflow*

Phaseflow [20] is a Python module maintained openly on GitHub. Phaseflow implements the methods of this paper using the open-source finite element library FEniCS [2]. The interface allows users to run these methods for the variational form (9) with any set of similarity parameters Ste, Pr, buoyancy model $\mathbf{f}_B(T)$, regularization $\phi(T)$, liquid and solid viscosities μ_L and μ_S, time step size Δt, and stabilization coefficient γ. The initial values (6) are set with a function in the solution space. Combined with the built-in FEniCS method, this allows the user to interpolate general mathematical expressions written in the C syntax, or to use an existing solution. Dirichlet boundary conditions (7) are also set with general C mathematical expressions. Additionally the interface allows users to control algorithm parameters, such as tolerances for the solvers, the Newton relaxation parameter ω, and other options.

The interface accepts any FEniCS mesh object, which can either be generated from scratch or can be converted from a standard format via the FEniCS library. Leveraging the abstract interfaces of FEniCS, Phaseflow is dimension-independent, and has been applied to 1D, 2D, and 3D spatial domains. In this work, unit square and rectangular prism domains have been used. Phaseflow writes all solutions to disk in the efficient XDMF+HDF5 format using the built-in FEniCS interface. Furthermore, this is combined with the H5Py library to write checkpoint files for easily restarting simulations at later times. Phaseflow has other ease-of-use features, such as a stopping criterion for steady state solutions. This was useful for the heat-driven cavity test in Sect. 5.2, where the Newton method required a small time step size.

4.3 *Reproducibility with Docker*

Phaseflow leverages open technology, and it is meant to contribute further to open scientific and engineering research. Open research in the computational sciences is often hindered by the difficulty of reproducing results, primarily because of the ever-increasing complexity of computing environments [6]. Recently, the software Docker has emerged as a technology which greatly facilitates reproducibility in the computational sciences.

Built upon an official FEniCS Docker image, we provide a Phaseflow Docker with all dependencies. Enabled by the Docker image and Phaseflow's test suite, changes to the master branch on GitHub are continuously tested using the Travis-CI continuous integration service [6]. Docker has been central to the development of Phaseflow. Initially, a primary motivation for its use was the existence of Docker images where most of the dependencies of Phaseflow are already installed and maintained by the FEniCS developers. This allows us to spend more time implementing models, and less time building tool chains. From there, maintaining a Docker image which runs Phaseflow takes little effort. With the Phaseflow Docker image in hand, we were then able to set up a continuous integration process with remarkably little effort.

The vast majority of Phaseflow's source code is covered with a suite of unit tests and integration tests written with PyTest. Most results in this paper are included in Phaseflow's test suite, while the others are covered by the example scripts and notebooks in its repository [20]. Such results are easy to reproduce, and we highly encourage the reader to do this. To further aid reproducibility, a specific version of Phaseflow has been archived (see the release versions at [20]) to coincide with this publication.

5 Verification

To verify our implementation, we consider a series of examples, including the lid-driven cavity benchmark from [12] extended with a solid subdomain, the heat-driven cavity benchmark from [19], an approximation of the analytical 1D Stefan problem from [1] with and without AMR, a preliminary 2D convection-coupled melting demonstration with AMR, and a preliminary 3D convection-coupled melting demonstration without AMR. Furthermore, for the Stefan problem, we verify the convergence orders of the temporal and spatial discretizations, and of the Newton method. Phaseflow's flexible interface described in Sect. 4 allows us to implement each of these applications with minimal effort, and to test the same lines of source code, increasing our confidence with the implementation.

5.1 Lid-Driven Cavity with Solid Subdomain

To test phase-dependent viscosity, we consider an extension of the lid-driven cavity benchmark based on data published in [12]. The standard benchmark uses a unit square geometry. We extend the geometry below the bottom wall, and set the temperature in the new region below the freezing temperature, such that the liquid subdomain per (4) covers the original unit square geometry. Therefore the initial temperature values are

$$T_0(\mathbf{x}) = \begin{cases} T_c & \text{for } y \leq 0, \\ T_h & \text{otherwise} \end{cases} \tag{17}$$

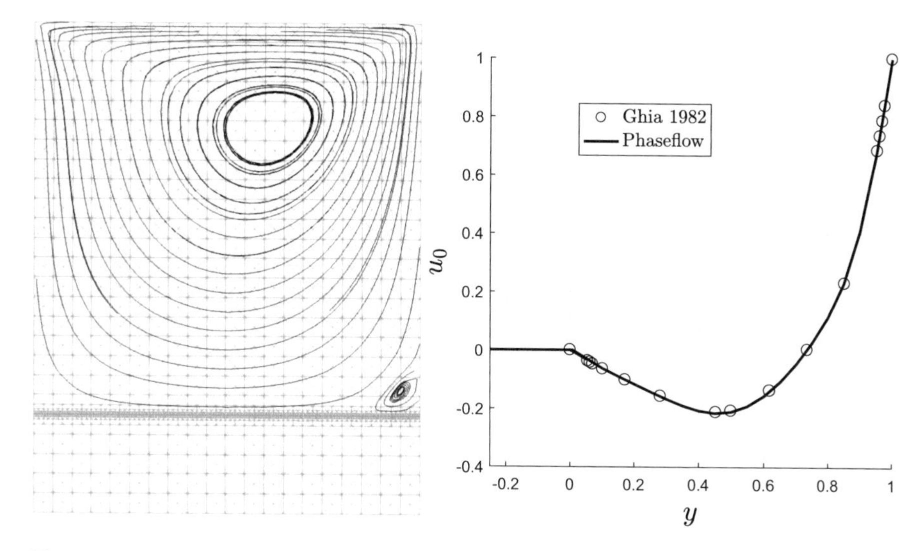

Fig. 3 Left: Result from lid-driven cavity test (Re = 100). The mesh (which is locally refined near the PCI) is shown in translucent gray. Velocity streamlines are shown in black. As expected, the moving lid causes circulation in the cavity, and we see the expected recirculation zone in the bottom-right corner of the liquid subdomain. Right: Horizontal velocity sampled from the vertical centerline, compared to benchmark data published in [12]

where we chose $T_h = 1$ and $T_c = -1$. For the semi-phase-field (4) we set central temperature $T_r = -0.01$ and smoothing parameter $r = 0.01$. We set variable viscosity (5) with $\mu_L = 1$ and $\mu_S = 10^8$. To capture this strong variation, we refined all cells of the initial mesh which touched coordinate $y = 0$ for four refinement cycles. Note that in this case, the local refinement is not adaptive. This problem at steady state fits our general model (9) with null buoyancy $\mathbf{f}_B = \mathbf{0}$ and with an arbitrarily large Prandtl number to nullify thermal conduction.

As boundary conditions $\mathbf{u}_D$, we set positive horizontal velocity on the lid, zero horizontal velocity away from the lid, and zero vertical velocity everywhere. We set initial velocity $\mathbf{u}_0$ similarly, which serves as a suitable initialization for solving until steady state with Newton's method. Because Phaseflow implements only the unsteady problem, we solve a single large time step to obtain the approximate steady state solution. It is advantageous to model the steady problem this way, when the goal is to verify components of our unsteady implementation. Figure 3 shows and discusses the result. The solution agrees very well with the published benchmark data.

5.2 *Heat-Driven Cavity*

To verify the coupled energy equation without phase-change, we compare to the heat-driven cavity benchmark data published in [19]. This problem fits our general model with constant viscosity $\mu_S = \mu_L = 1$ and arbitrarily large Stefan number. We handle buoyancy with an idealized linear[1] Boussinesq model

$$\mathbf{f}_B(T) = \frac{\mathrm{Ra}}{\mathrm{Pr}\mathrm{Re}^2} T \begin{pmatrix} 0 \\ -1 \end{pmatrix}, \tag{18}$$

The momentum equation (2) is scaled such that the Reynolds number Re is always unity. For this benchmark, the Rayleigh number is $\mathrm{Ra} = 10^6$ and the Prandtl number is $\mathrm{Pr} = 0.71$. This Rayleigh number is considered to be high, and demonstrates substantial natural convection. We set homogeneous Dirichlet (i.e. no slip) boundary conditions on the velocity $\mathbf{u}_D = \mathbf{0}$, and non-homogeneous Dirichlet boundary conditions for the temperature with hot and cool temperatures, $T_h = 0.5$ and $T_c = -0.5$, respectively on the left and right vertical walls, i.e.

$$T_D(\mathbf{x}) = \begin{cases} T_h & \text{for } x = 0, \\ T_c & \text{for } x = 1 \end{cases} \tag{19}$$

Again we solve the unsteady problem until it reaches steady state. For this we initialize the velocity field to zero (i.e. $\mathbf{u}_0 = \mathbf{0}$) and the temperature field to vary linearly between the hot and cold walls (i.e. $T_0(\mathbf{x}) = T_h + x(T_c - T_h)$). Unlike with the lid-driven cavity, here we cannot obtain the steady state solution in a single time step, because the initial guess is not sufficient for the Newton method to converge. Therefore we solve a sequence of time steps using $\Delta t = 0.001$. Figure 4 shows the successful result with further discussion.

5.3 *The Stefan Problem*

To verify the energy equation with phase-change, we compare to the analytical 1D Stefan problem as written in [1], with parameters comparable to the octadecane PCM melting benchmark in Sect. 5.4, including the Stefan number $\mathrm{Ste} = 0.045$. This problem fits our general model with nullified buoyancy $f_B = 0$, unity Prandtl number, and zero initial velocity $u_0 = 0$ (which remains zero). Solving the Stefan problem as a special case of the coupled problem results in an unnecessarily large system. In this case, the vast majority of degrees of freedom are trivial. From an implementation perspective, this exercise is quite valuable, because we test the exact lines of code which are used for the coupled problem.

[1]The method handles nonlinear buoyancy with (13). Phaseflow's test suite includes a benchmark with the nonlinear density anomaly of water.

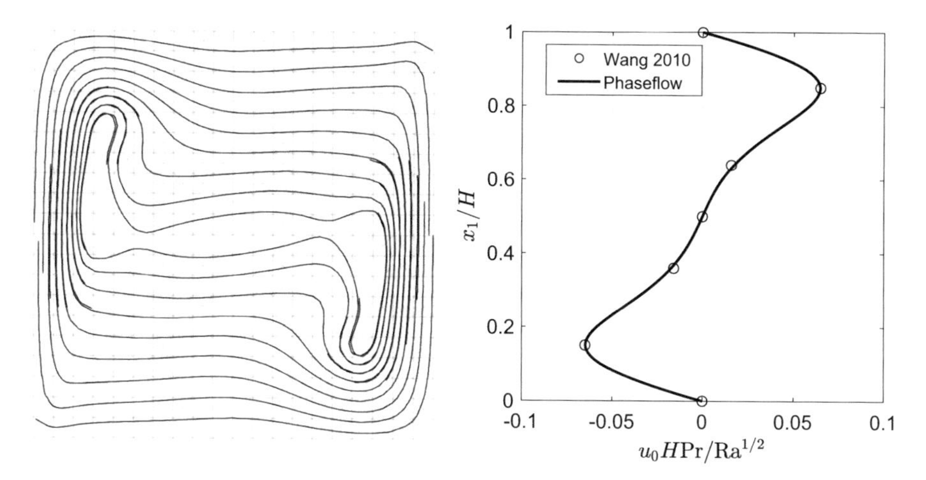

Fig. 4 Left: Result from heat-driven cavity test ($\mathrm{Ra} = 10^6$). The mesh is shown in translucent gray. Velocity streamlines are shown in black. As expected, we see the flow circulating, being forced upward near the hot wall and downward near the cold wall. Right: The horizontal component of velocity is plotted along the vertical centerline. The solution agrees very well with the data published in [19]

We again set temperature boundary conditions (19), this time with $T_h = 1$ and $T_c = -0.01$. The initial temperature field is set such that a thin layer of melt exists near the hot wall, with the rest of the domain at the cold wall temperature, i.e.

$$T_0(x) = \begin{cases} T_h & \text{for } x \leq x_0^*, \\ T_c & \text{otherwise} \end{cases} \tag{20}$$

We parameterize this initial PCI position as

$$x_0^* = \frac{L}{N_0} 2^{1-q} \tag{21}$$

where q is the number of initial hot wall refinement cycles. This ensures that for the given initial mesh, the thinnest possible layer of melt exists. For the semi-phase-field regularization (4) we set parameters $T_r = 0, r = 0.01$.

We simulate until time $t_f = 0.1$, using the time step size $\Delta t = 0.001$. In testing, this size was needed to bound the point-wise error between the numerical and analytical solutions from above by $T(x) - T_{\text{exact}}(x) < 0.01 \quad \forall x \in \{0, 0.025, 0.05, 0.075, 0.1, 0.5, 1\}$ at time $t = t_f$ (shown in Fig. 5). This discrete set of verification points is a good substitute for the global solution. Figure 5 shows the successful result and further discussion.

We also take this opportunity to demonstrate AMR. We set the adaptive goal (16) with tolerance $\epsilon_M = 10^{-6}$. During the time-dependent simulation, new cells are only added, and never removed, i.e. the mesh is never coarsened. This is an

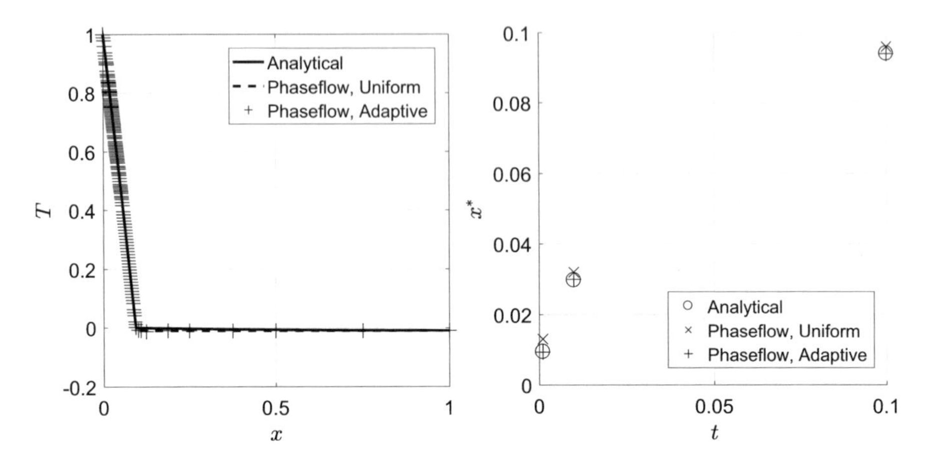

Fig. 5 Left: Comparison at $t = 0.1$ between the analytical Stefan problem solution, Phaseflow using a uniform mesh of 311 cells (For the uniform case, this is the minimum number of cells for which the Newton method would converge. Similarly for the AMR case, starting with a coarser initial mesh disrupts the Newton method.) and Phaseflow using AMR (with 133 cells in the final adapted mesh). The AMR solution is shown with a marker at every mesh vertex. This emphasizes the much smaller cell sizes in the wake of the PCI. Right: Comparison of the PCI positions at three times between the analytical, uniform, and AMR solutions. We observe a bias in the uniform mesh solution which shrinks over time. This is primarily because of the mesh-dependent initial PCI position (21)

unfortunate limitation of the current AMR algorithm. We see that cells have been clustered near the PCI, and these clusters remain everywhere the PCI has been during the time-dependent simulation. Rightward of the PCI, the cells grow much larger, e.g. the right half of the domain is covered by only two cells. Even without coarsening, already in 1D this is a large improvement. The gains in 2D and 3D will be even greater.

5.4 2D Convection-Coupled Melting of Octadecane PCM

To demonstrate the entire coupled system, we present a preliminary result for the convection-coupled melting of the octadecane PCM benchmark from [19] and [8]. This problem uses all aspects of our general model (9). We set variable viscosity (5) (with $\mu_L = 1$ and $\mu_S = 10^8$), the buoyancy model (18) (with $\mathrm{Ra} = 3.27 \times 10^5$, $\mathrm{Pr} = 56.2$, and $\mathrm{Re} = 1$), no-slip velocity boundary conditions $\mathbf{u}_D = \mathbf{0}$, and again the temperature boundary conditions (19) with $T_h = 1$ and $T_c = -0.01$. Again we initialize the temperature field with (20) such that the simulation begins with a thin layer of melt, and initialize a stationary velocity field $\mathbf{u}_0 = \mathbf{0}$. For the semi-phase-field (4), we set regularization parameters $T_r = 0.01$ and $r = 0.025$. For the time-dependent simulation, we set $\Delta t = 1$. For AMR, we again set the adaptive

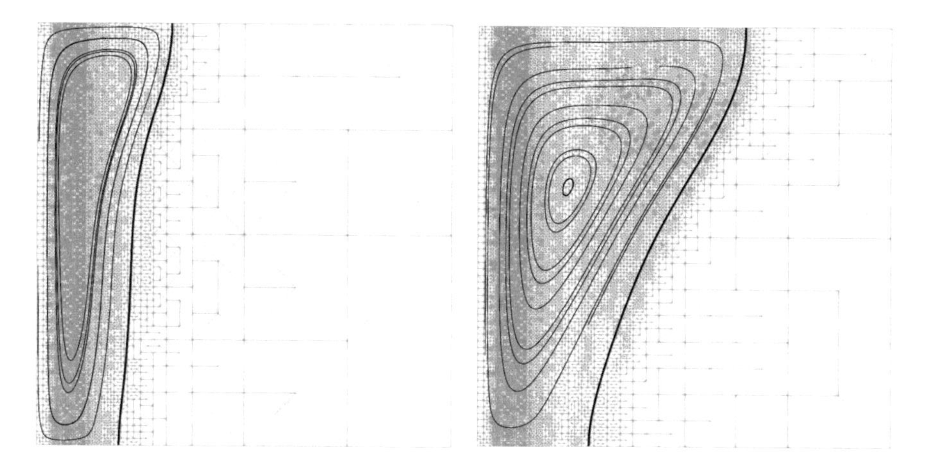

Fig. 6 Preliminary result for the 2D convection-coupled octadecane PCM melting benchmark. Solutions are shown at simulated times $t = 36$ and $t = 80$ (on the left and right). The mesh is shown in translucent gray to highlight the adaptive mesh refinement (AMR). The PCI is shown as a thick black temperature isoline. Velocity streamlines are shown as thin black lines. This result looks promising when compared to octadecane PCM melting results from [19] and [8]. We see the flow circulating, and the PCI advanced more quickly at the top than at the bottom. We also see that the lack of grid coarsening is becoming expensive. By the final time, many refined cells from earlier times are likely not needed

goal (16) from our Stefan problem example, but with tolerance $\epsilon_M = 10^{-5}$ until $t = 36$ and $\epsilon_M = 0.5 \times 10^{-5}$ after that time.

The result in Fig. 6 is promising; but compared to the benchmark, we see that the PCI has not advanced far enough by simulated time $t = 80$. We still need to investigate the effects of our choices for Δt, r, x_0^*, $M(\mathbf{w})$, ϵ_M, and the initial mesh refinement. Of these, we know that a much smaller smoothing parameter r was used in [8].

5.5 3D Convection-Coupled Melting

The dimension-independent implementation, facilitated by the abstractions from FEniCS, allow us to quickly demonstrate a 3D example. We consider the a problem similar to the 2D convection-coupled melting in Sect. 5.4, but with the domain extruded in the z direction, adiabatic no-slip boundary conditions on the walls parallel to the z plane, similarity parameters $\text{Ste} = 1$, $\text{Ra} = 10^6$, $\text{Pr} = 0.71$, and numerical parameters $\mu_S = 10^4$, $r = 0.05$. For the previous examples, we employed the "full" Newton method with $\omega = 1$. For this problem, we relaxed the Newton method with a factor of $\omega = 0.8$. Figure 7 shows the preliminary result.

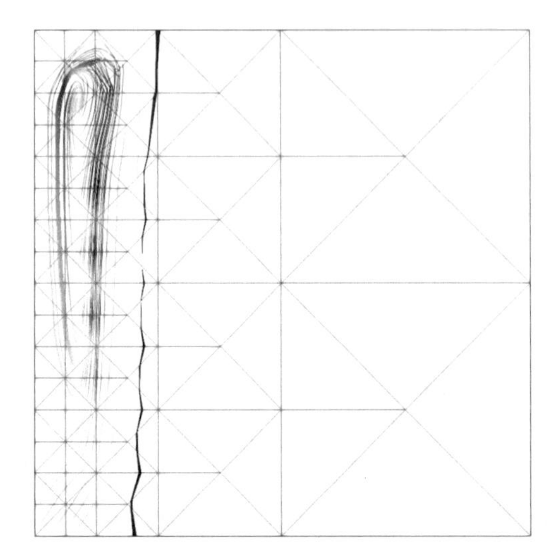
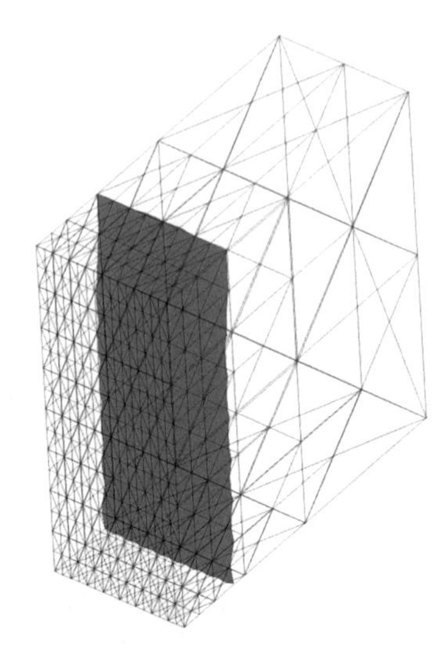

Fig. 7 Preliminary result for 3D convection-coupled melting, with front and isometric views (on the left and right). The mesh is shown in translucent gray to highlight the local mesh refinement. The PCI is a darker translucent gray iso-surface. The streamlines are colored by a grayscale from white to black as velocity magnitude increases. We use a coarse mesh and omit AMR, because the refined 3D problem quickly exceeded the capabilities of the desktop-scale computers used for this work. Despite the coarse mesh, we still observe the primary features we expect in the solution. The flow is circulating, and we see that the top of the PCI advancing more quickly than the bottom

5.6 Convergence

To verify the accuracies of the finite difference time discretization, finite element space discretization, and Newton linearization methods, we consider the 1D Stefan problem from Sect. 5.3. The mixed finite element formulation was shown to be second order accurate for the incompressible Navier-Stokes equations in [7]. We are not aware of such a result for the energy-coupled problem. Here, as a first step, we focus only on the energy equation with phase-change.

Based on the choices of discretizations, with fully implicit Euler for time and finite elements for space, we expect first order convergence in time and second order in space. Figure 8 compares this with the actual convergence orders of Phaseflow's solution from Sect. 5.3. With respect to the time step sizes Δt, the observed convergence order is only slightly higher than expected. With respect to the grid spacing h, for a sufficiently smooth solution with well-behaved data, we should expect second order accuracy. Though there is some deviation from second order near $h = 0.001$, the results in Fig. 8 show good agreement.

Next, we verify the accuracy of the Newton method. Figure 8 shows its convergence behavior for multiple time steps, highlighting the difference in performance between the first and last time steps. Ideally, Newton's method will converge quadratically for well-posed problems with a suitable initial guess. At earlier times,

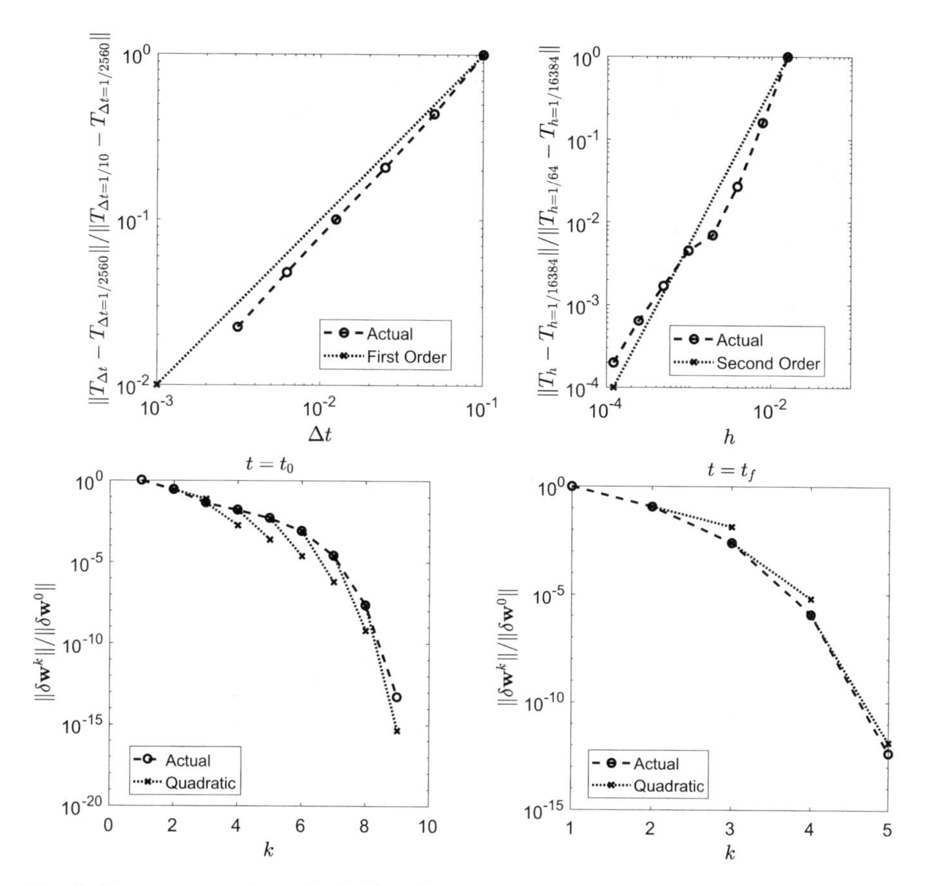

Fig. 8 Convergence study results for Phaseflow applied to the Stefan problem from Sect. 5.3. Top: Temporal (on the left) and spatial (on the right) convergence results based on the solution at the final time t_f. Bottom: Newton method convergence shown for multiple time steps. Each sub-plot has a series of lines which represent ideal (quadratic) iterations

a few Newton iterations pass before reaching approximately quadratic convergence. Later times converge slightly better than quadratically.

6 Conclusions and Outlook

In this work, we presented the computational tool Phaseflow for simulating the convection-coupled melting and solidification of PCM's. Our work is based on the numerical method proposed in [8], and so we used an enthalpy formulated, single-domain semi-phase-field, variable viscosity, finite element method, with monolithic system coupling and global Newton linearization. The primary difference between our numerical approach and that in [8] is the mesh adaptivity algorithm, where we employ the dual-weighted residual method for goal-oriented AMR. We implemented the method into our open source Python module Phaseflow, based on the finite element software library FEniCS, and verified this against a series of classical benchmarks. We obtained a promising result for the octadecane PCM convection-coupled melting benchmark. Leveraging our dimension-independent implementation, we contributed detailed convergence results for the 1D phase-change problem, and have applied the method to a preliminary 3D convection-coupled melting example. Furthermore, we openly shared a Docker container which allows anyone to reproduce our results in the same software environment. It is our hope that this facilitates the further development of this and related methods. The method appears promising, and Phaseflow is ready for application to interesting problems.

FEniCS was a good choice as the base of our implementation, allowing us to focus on the models and numerical methods rather than on implementing standard algorithms. Furthermore, the existing Docker software container allowed us to quickly leverage the library. Some difficulties do remain, primarily: (1) FEniCS lacks mesh coarsening capability, prohibiting the efficient application of AMR to the moving PCI problem. (2) The adaptive solver (using dual-weighted residual goal-oriented AMR) has not been implemented for distributed memory systems, therefore currently limiting the implementation's applicability to a single compute node, prohibiting realistic 3D applications. On the other hand, there is yet potential to simplify Phaseflow with existing FEniCS features. Most interestingly, there is an automatic differentiation capability which could serve as an alternative to computing the Gâteaux derivative (13) directly. During the development of Phaseflow, we have successfully applied this feature to some simplified cases; and we would like to further explore this capability.

Next steps of our work include the development of robustness features, including... (1) a priori bounds on the time step size Δt, based on the properties of the linear system (12) and initial values (6), (2) an algorithm for obtaining the initial mesh on which to obtain the first solution to begin AMR. The latter is important, because we discovered a limitation with relying on AMR to resolve strong nonlinearities. Typically, goal-oriented AMR is used to provide either the most accurate solution (with respect to the goal) for a given cost, or the minimum-cost discretization for a given accuracy (with respect to the goal). In the case of the presented method, AMR is required for the Newton method to converge and to hence obtain the first solution in the hierarchy. Handling these issues is necessary before practically applying this method to a wider range of realistic problems.

References

1. Alexiades, V.: Mathematical Modeling of Melting and Freezing Processes. CRC Press, West Palm Beach (1992)
2. Alnæs, M., Blechta, J., Hake, J., et al.: The fenics project version 1.5. Arch. Numer. Softw. **3**(100), 9–23 (2015)

3. Bangerth, W., Rannacher, R.: Adaptive Finite Element Methods for Differential Equations. Springer Science & Business Media, New York (2003)
4. Batchelor, G.K., Moffatt, H.K., Worster, M.G., Osborn, T.R.: Perspectives in Fluid Dynamics. Cambridge University Press, Cambridge (2000)
5. Belhamadia, Y., Fortin, A., Chamberland, É.: Anisotropic mesh adaptation for the solution of the stefan problem. J. Comput. Phys. **194**(1), 233–255 (2004)
6. Boettiger, C.: An introduction to docker for reproducible research. ACM SIGOPS Oper. Syst. Rev. **49**(1), 71–79 (2015)
7. Brezzi, F., Fortin, M.: Mixed and Hybrid Finite Element Methods, vol. 15. Springer Science & Business Media, New York (1991)
8. Danaila, I., Moglan, R., Hecht, F., Le Masson, S.: A newton method with adaptive finite elements for solving phase-change problems with natural convection. J. Comput. Phys. **274**, 826–840 (2014)
9. Dinniman, M.S., Asay-Davis, X.S., Galton-Fenzi, B.K., et al.: Modeling ice shelf/ocean interaction in antarctica: a review. Oceanography **29**(4), 144–153 (2016)
10. Donea, J., Huerta, A.: Finite Element Methods for Flow Problems. Wiley, London (2003)
11. Dutil, Y., Rousse, D.R., Salah, N.B., et al.: A review on phase-change materials: mathematical modeling and simulations. Renew. Sust. Energ. Rev. **15**(1), 112–130 (2011)
12. Ghia, U.K.N.G., Ghia, K.N., Shin, K.N.: High-Re solutions for incompressible flow using the Navier-Stokes equations and a multigrid method. J. Comput. Phys. **48**(3), 387–411 (1982)
13. Hsu, H.W., Postberg, F., Sekine, Y., et al.: Ongoing hydrothermal activities within enceladus. Nature **519**(7542), 207–210 (2015)
14. Kelly, D.W., De SR Gago, J.P., Zienkiewicz, O.C., et al.: A posteriori error analysis and adaptive processes in the finite element method: part I—error analysis. Int. J. Numer. Methods Eng. **19**(11), 1593–1619 (1983)
15. Kowalski, J., Linder, P., Zierke, S., et al.: Navigation technology for exploration of glacier ice with maneuverable melting probes. Cold Reg. Sci. Technol. **123**, 53–70 (2016)
16. Schüller, K., Berkels, B., Kowalski, J.: Integrated modeling and validation for phase change with natural convection. (2018). https://arxiv.org/abs/1801.03699
17. Voller, V.R., Cross, M., Markatos, N.C.: An enthalpy method for convection/diffusion phase change. Int. J. Numer. Methods Eng. **24**(1), 271–284 (1987)
18. Voller, V.R., Swaminathan, C.R., Thomas, B.G.: Fixed grid techniques for phase change problems: a review. Int. J. Numer. Methods Eng. **30**(4), 875–898 (1990)
19. Wang, S., Faghri, A., Bergman, T.L.: A comprehensive numerical model for melting with natural convection. Int. J. Heat Mass Transf. **53**(9), 1986–2000 (2010)
20. Zimmerman, A.: Phaseflow (Open Source Python module). https://github.com/geo-fluid-dynamics/phaseflow-fenics

Author Index

M. Schäfer et al. (eds.), *Recent Advances in Computational Engineering*,
Lecture Notes in Computational Science and Engineering 124,
https://doi.org/10.1007/978-3-319-93891-2

Editorial Policy

1. Volumes in the following three categories will be published in LNCSE:

i) Research monographs
ii) Tutorials
iii) Conference proceedings

Those considering a book which might be suitable for the series are strongly advised to contact the publisher or the series editors at an early stage.

2. Categories i) and ii). Tutorials are lecture notes typically arising via summer schools or similar events, which are used to teach graduate students. These categories will be emphasized by Lecture Notes in Computational Science and Engineering. **Submissions by interdisciplinary teams of authors are encouraged.** The goal is to report new developments – quickly, informally, and in a way that will make them accessible to non-specialists. In the evaluation of submissions timeliness of the work is an important criterion. Texts should be well-rounded, well-written and reasonably self-contained. In most cases the work will contain results of others as well as those of the author(s). In each case the author(s) should provide sufficient motivation, examples, and applications. In this respect, Ph.D. theses will usually be deemed unsuitable for the Lecture Notes series. Proposals for volumes in these categories should be submitted either to one of the series editors or to Springer-Verlag, Heidelberg, and will be refereed. A provisional judgement on the acceptability of a project can be based on partial information about the work: a detailed outline describing the contents of each chapter, the estimated length, a bibliography, and one or two sample chapters – or a first draft. A final decision whether to accept will rest on an evaluation of the completed work which should include

- at least 100 pages of text;
- a table of contents;
- an informative introduction perhaps with some historical remarks which should be accessible to readers unfamiliar with the topic treated;
- a subject index.

3. Category iii). Conference proceedings will be considered for publication provided that they are both of exceptional interest and devoted to a single topic. One (or more) expert participants will act as the scientific editor(s) of the volume. They select the papers which are suitable for inclusion and have them individually refereed as for a journal. Papers not closely related to the central topic are to be excluded. Organizers should contact the Editor for CSE at Springer at the planning stage, see *Addresses* below.

In exceptional cases some other multi-author-volumes may be considered in this category.

4. Only works in English will be considered. For evaluation purposes, manuscripts may be submitted in print or electronic form, in the latter case, preferably as pdf- or zipped ps-files. Authors are requested to use the LaTeX style files available from Springer at http://www.springer.com/gp/authors-editors/book-authors-editors/manuscript-preparation/5636 (Click on LaTeX Template → monographs or contributed books).

For categories ii) and iii) we strongly recommend that all contributions in a volume be written in the same LaTeX version, preferably LaTeX2e. Electronic material can be included if appropriate. Please contact the publisher.

Careful preparation of the manuscripts will help keep production time short besides ensuring satisfactory appearance of the finished book in print and online.

5. The following terms and conditions hold. Categories i), ii) and iii):

Authors receive 50 free copies of their book. No royalty is paid.
Volume editors receive a total of 50 free copies of their volume to be shared with authors, but no royalties.

Authors and volume editors are entitled to a discount of 40 % on the price of Springer books purchased for their personal use, if ordering directly from Springer.

6. Springer secures the copyright for each volume.

Addresses:

Timothy J. Barth
NASA Ames Research Center
NAS Division
Moffett Field, CA 94035, USA
barth@nas.nasa.gov

Michael Griebel
Institut für Numerische Simulation
der Universität Bonn
Wegelerstr. 6
53115 Bonn, Germany
griebel@ins.uni-bonn.de

David E. Keyes
Mathematical and Computer Sciences
and Engineering
King Abdullah University of Science
and Technology
P.O. Box 55455
Jeddah 21534, Saudi Arabia
david.keyes@kaust.edu.sa

and

Department of Applied Physics
and Applied Mathematics
Columbia University
500 W. 120 th Street
New York, NY 10027, USA
kd2112@columbia.edu

Risto M. Nieminen
Department of Applied Physics
Aalto University School of Science
and Technology
00076 Aalto, Finland
risto.nieminen@aalto.fi

Dirk Roose
Department of Computer Science
Katholieke Universiteit Leuven
Celestijnenlaan 200A
3001 Leuven-Heverlee, Belgium
dirk.roose@cs.kuleuven.be

Tamar Schlick
Department of Chemistry
and Courant Institute
of Mathematical Sciences
New York University
251 Mercer Street
New York, NY 10012, USA
schlick@nyu.edu

Editor for Computational Science
and Engineering at Springer:
Jan Holland
Springer-Verlag
Mathematics Editorial
Tiergartenstrasse 17
69121 Heidelberg, Germany
jan.holland@springer.com

Lecture Notes in Computational Science and Engineering

1. D. Funaro, *Spectral Elements for Transport-Dominated Equations.*
2. H.P. Langtangen, *Computational Partial Differential Equations.* Numerical Methods and Diffpack Programming.
3. W. Hackbusch, G. Wittum (eds.), *Multigrid Methods V.*
4. P. Deuflhard, J. Hermans, B. Leimkuhler, A.E. Mark, S. Reich, R.D. Skeel (eds.), *Computational Molecular Dynamics: Challenges, Methods, Ideas.*
5. D. Kröner, M. Ohlberger, C. Rohde (eds.), *An Introduction to Recent Developments in Theory and Numerics for Conservation Laws.*
6. S. Turek, *Efficient Solvers for Incompressible Flow Problems.* An Algorithmic and Computational Approach.
7. R. von Schwerin, ***Multi Body System SIMulation.*** Numerical Methods, Algorithms, and Software.
8. H.-J. Bungartz, F. Durst, C. Zenger (eds.), *High Performance Scientific and Engineering Computing.*
9. T.J. Barth, H. Deconinck (eds.), *High-Order Methods for Computational Physics.*
10. H.P. Langtangen, A.M. Bruaset, E. Quak (eds.), *Advances in Software Tools for Scientific Computing.*
11. B. Cockburn, G.E. Karniadakis, C.-W. Shu (eds.), *Discontinuous Galerkin Methods.* Theory, Computation and Applications.
12. U. van Rienen, *Numerical Methods in Computational Electrodynamics.* Linear Systems in Practical Applications.
13. B. Engquist, L. Johnsson, M. Hammill, F. Short (eds.), *Simulation and Visualization on the Grid.*
14. E. Dick, K. Riemslagh, J. Vierendeels (eds.), *Multigrid Methods VI.*
15. A. Frommer, T. Lippert, B. Medeke, K. Schilling (eds.), *Numerical Challenges in Lattice Quantum Chromodynamics.*
16. J. Lang, *Adaptive Multilevel Solution of Nonlinear Parabolic PDE Systems.* Theory, Algorithm, and Applications.
17. B.I. Wohlmuth, *Discretization Methods and Iterative Solvers Based on Domain Decomposition.*
18. U. van Rienen, M. Günther, D. Hecht (eds.), *Scientific Computing in Electrical Engineering.*
19. I. Babuška, P.G. Ciarlet, T. Miyoshi (eds.), *Mathematical Modeling and Numerical Simulation in Continuum Mechanics.*
20. T.J. Barth, T. Chan, R. Haimes (eds.), *Multiscale and Multiresolution Methods.* Theory and Applications.
21. M. Breuer, F. Durst, C. Zenger (eds.), *High Performance Scientific and Engineering Computing.*
22. K. Urban, *Wavelets in Numerical Simulation.* Problem Adapted Construction and Applications.
23. L.F. Pavarino, A. Toselli (eds.), *Recent Developments in Domain Decomposition Methods.*

24. T. Schlick, H.H. Gan (eds.), *Computational Methods for Macromolecules: Challenges and Applications.*

25. T.J. Barth, H. Deconinck (eds.), *Error Estimation and Adaptive Discretization Methods in Computational Fluid Dynamics.*

26. M. Griebel, M.A. Schweitzer (eds.), *Meshfree Methods for Partial Differential Equations.*

27. S. Müller, *Adaptive Multiscale Schemes for Conservation Laws.*

28. C. Carstensen, S. Funken, W. Hackbusch, R.H.W. Hoppe, P. Monk (eds.), *Computational Electromagnetics.*

29. M.A. Schweitzer, *A Parallel Multilevel Partition of Unity Method for Elliptic Partial Differential Equations.*

30. T. Biegler, O. Ghattas, M. Heinkenschloss, B. van Bloemen Waanders (eds.), *Large-Scale PDE-Constrained Optimization.*

31. M. Ainsworth, P. Davies, D. Duncan, P. Martin, B. Rynne (eds.), *Topics in Computational Wave Propagation.* Direct and Inverse Problems.

32. H. Emmerich, B. Nestler, M. Schreckenberg (eds.), *Interface and Transport Dynamics.* Computational Modelling.

33. H.P. Langtangen, A. Tveito (eds.), *Advanced Topics in Computational Partial Differential Equations.* Numerical Methods and Diffpack Programming.

34. V. John, *Large Eddy Simulation of Turbulent Incompressible Flows.* Analytical and Numerical Results for a Class of LES Models.

35. E. Bänsch (ed.), *Challenges in Scientific Computing - CISC 2002.*

36. B.N. Khoromskij, G. Wittum, *Numerical Solution of Elliptic Differential Equations by Reduction to the Interface.*

37. A. Iske, *Multiresolution Methods in Scattered Data Modelling.*

38. S.-I. Niculescu, K. Gu (eds.), *Advances in Time-Delay Systems.*

39. S. Attinger, P. Koumoutsakos (eds.), *Multiscale Modelling and Simulation.*

40. R. Kornhuber, R. Hoppe, J. Périaux, O. Pironneau, O. Wildlund, J. Xu (eds.), *Domain Decomposition Methods in Science and Engineering.*

41. T. Plewa, T. Linde, V.G. Weirs (eds.), *Adaptive Mesh Refinement – Theory and Applications.*

42. A. Schmidt, K.G. Siebert, *Design of Adaptive Finite Element Software.* The Finite Element Toolbox ALBERTA.

43. M. Griebel, M.A. Schweitzer (eds.), *Meshfree Methods for Partial Differential Equations II.*

44. B. Engquist, P. Lötstedt, O. Runborg (eds.), *Multiscale Methods in Science and Engineering.*

45. P. Benner, V. Mehrmann, D.C. Sorensen (eds.), *Dimension Reduction of Large-Scale Systems.*

46. D. Kressner, *Numerical Methods for General and Structured Eigenvalue Problems.*

47. A. Boriçi, A. Frommer, B. Joó, A. Kennedy, B. Pendleton (eds.), *QCD and Numerical Analysis III.*

48. F. Graziani (ed.), *Computational Methods in Transport.*

49. B. Leimkuhler, C. Chipot, R. Elber, A. Laaksonen, A. Mark, T. Schlick, C. Schütte, R. Skeel (eds.), *New Algorithms for Macromolecular Simulation.*

50. M. Bücker, G. Corliss, P. Hovland, U. Naumann, B. Norris (eds.), *Automatic Differentiation: Applications, Theory, and Implementations.*

51. A.M. Bruaset, A. Tveito (eds.), *Numerical Solution of Partial Differential Equations on Parallel Computers.*

52. K.H. Hoffmann, A. Meyer (eds.), *Parallel Algorithms and Cluster Computing.*

53. H.-J. Bungartz, M. Schäfer (eds.), *Fluid-Structure Interaction.*

54. J. Behrens, *Adaptive Atmospheric Modeling.*

55. O. Widlund, D. Keyes (eds.), *Domain Decomposition Methods in Science and Engineering XVI.*

56. S. Kassinos, C. Langer, G. Iaccarino, P. Moin (eds.), *Complex Effects in Large Eddy Simulations.*

57. M. Griebel, M.A Schweitzer (eds.), *Meshfree Methods for Partial Differential Equations III.*

58. A.N. Gorban, B. Kégl, D.C. Wunsch, A. Zinovyev (eds.), *Principal Manifolds for Data Visualization and Dimension Reduction.*

59. H. Ammari (ed.), *Modeling and Computations in Electromagnetics: A Volume Dedicated to Jean-Claude Nédélec.*

60. U. Langer, M. Discacciati, D. Keyes, O. Widlund, W. Zulehner (eds.), *Domain Decomposition Methods in Science and Engineering XVII.*

61. T. Mathew, *Domain Decomposition Methods for the Numerical Solution of Partial Differential Equations.*

62. F. Graziani (ed.), *Computational Methods in Transport: Verification and Validation.*

63. M. Bebendorf, *Hierarchical Matrices.* A Means to Efficiently Solve Elliptic Boundary Value Problems.

64. C.H. Bischof, H.M. Bücker, P. Hovland, U. Naumann, J. Utke (eds.), *Advances in Automatic Differentiation.*

65. M. Griebel, M.A. Schweitzer (eds.), *Meshfree Methods for Partial Differential Equations IV.*

66. B. Engquist, P. Lötstedt, O. Runborg (eds.), *Multiscale Modeling and Simulation in Science.*

67. I.H. Tuncer, Ü. Gülcat, D.R. Emerson, K. Matsuno (eds.), *Parallel Computational Fluid Dynamics 2007.*

68. S. Yip, T. Diaz de la Rubia (eds.), *Scientific Modeling and Simulations.*

69. A. Hegarty, N. Kopteva, E. O'Riordan, M. Stynes (eds.), *BAIL* 2008 – *Boundary and Interior Layers.*

70. M. Bercovier, M.J. Gander, R. Kornhuber, O. Widlund (eds.), *Domain Decomposition Methods in Science and Engineering XVIII.*

71. B. Koren, C. Vuik (eds.), *Advanced Computational Methods in Science and Engineering.*

72. M. Peters (ed.), *Computational Fluid Dynamics for Sport Simulation.*

73. H.-J. Bungartz, M. Mehl, M. Schäfer (eds.), *Fluid Structure Interaction II - Modelling, Simulation, Optimization.*

74. D. Tromeur-Dervout, G. Brenner, D.R. Emerson, J. Erhel (eds.), *Parallel Computational Fluid Dynamics 2008.*

75. A.N. Gorban, D. Roose (eds.), *Coping with Complexity: Model Reduction and Data Analysis.*

76. J.S. Hesthaven, E.M. Rønquist (eds.), *Spectral and High Order Methods for Partial Differential Equations.*

77. M. Holtz, *Sparse Grid Quadrature in High Dimensions with Applications in Finance and Insurance.*

78. Y. Huang, R. Kornhuber, O.Widlund, J. Xu (eds.), *Domain Decomposition Methods in Science and Engineering XIX.*

79. M. Griebel, M.A. Schweitzer (eds.), *Meshfree Methods for Partial Differential Equations V.*

80. P.H. Lauritzen, C. Jablonowski, M.A. Taylor, R.D. Nair (eds.), *Numerical Techniques for Global Atmospheric Models.*

81. C. Clavero, J.L. Gracia, F.J. Lisbona (eds.), *BAIL 2010 – Boundary and Interior Layers, Computational and Asymptotic Methods.*

82. B. Engquist, O. Runborg, Y.R. Tsai (eds.), *Numerical Analysis and Multiscale Computations.*

83. I.G. Graham, T.Y. Hou, O. Lakkis, R. Scheichl (eds.), *Numerical Analysis of Multiscale Problems.*

84. A. Logg, K.-A. Mardal, G. Wells (eds.), *Automated Solution of Differential Equations by the Finite Element Method.*

85. J. Blowey, M. Jensen (eds.), *Frontiers in Numerical Analysis - Durham 2010.*

86. O. Kolditz, U.-J. Gorke, H. Shao, W. Wang (eds.), *Thermo-Hydro-Mechanical-Chemical Processes in Fractured Porous Media - Benchmarks and Examples.*

87. S. Forth, P. Hovland, E. Phipps, J. Utke, A. Walther (eds.), *Recent Advances in Algorithmic Differentiation.*

88. J. Garcke, M. Griebel (eds.), *Sparse Grids and Applications.*

89. M. Griebel, M.A. Schweitzer (eds.), *Meshfree Methods for Partial Differential Equations VI.*

90. C. Pechstein, *Finite and Boundary Element Tearing and Interconnecting Solvers for Multiscale Problems.*

91. R. Bank, M. Holst, O. Widlund, J. Xu (eds.), *Domain Decomposition Methods in Science and Engineering XX.*

92. H. Bijl, D. Lucor, S. Mishra, C. Schwab (eds.), *Uncertainty Quantification in Computational Fluid Dynamics.*

93. M. Bader, H.-J. Bungartz, T. Weinzierl (eds.), *Advanced Computing.*

94. M. Ehrhardt, T. Koprucki (eds.), *Advanced Mathematical Models and Numerical Techniques for Multi-Band Effective Mass Approximations.*

95. M. Azaïez, H. El Fekih, J.S. Hesthaven (eds.), *Spectral and High Order Methods for Partial Differential Equations ICOSAHOM 2012.*

96. F. Graziani, M.P. Desjarlais, R. Redmer, S.B. Trickey (eds.), *Frontiers and Challenges in Warm Dense Matter.*

97. J. Garcke, D. Pflüger (eds.), *Sparse Grids and Applications – Munich 2012.*

98. J. Erhel, M. Gander, L. Halpern, G. Pichot, T. Sassi, O. Widlund (eds.), *Domain Decomposition Methods in Science and Engineering XXI.*

99. R. Abgrall, H. Beaugendre, P.M. Congedo, C. Dobrzynski, V. Perrier, M. Ricchiuto (eds.), *High Order Nonlinear Numerical Methods for Evolutionary PDEs - HONOM 2013.*

100. M. Griebel, M.A. Schweitzer (eds.), *Meshfree Methods for Partial Differential Equations VII.*

101. R. Hoppe (ed.), *Optimization with PDE Constraints - OPTPDE 2014.*

102. S. Dahlke, W. Dahmen, M. Griebel, W. Hackbusch, K. Ritter, R. Schneider, C. Schwab, H. Yserentant (eds.), *Extraction of Quantifiable Information from Complex Systems.*

103. A. Abdulle, S. Deparis, D. Kressner, F. Nobile, M. Picasso (eds.), *Numerical Mathematics and Advanced Applications - ENUMATH 2013.*

104. T. Dickopf, M.J. Gander, L. Halpern, R. Krause, L.F. Pavarino (eds.), *Domain Decomposition Methods in Science and Engineering XXII.*

105. M. Mehl, M. Bischoff, M. Schäfer (eds.), *Recent Trends in Computational Engineering - CE2014.* Optimization, Uncertainty, Parallel Algorithms, Coupled and Complex Problems.

106. R.M. Kirby, M. Berzins, J.S. Hesthaven (eds.), *Spectral and High Order Methods for Partial Differential Equations - ICOSAHOM'14.*

107. B. Jüttler, B. Simeon (eds.), *Isogeometric Analysis and Applications 2014.*

108. P. Knobloch (ed.), *Boundary and Interior Layers, Computational and Asymptotic Methods – BAIL 2014.*

109. J. Garcke, D. Pflüger (eds.), *Sparse Grids and Applications – Stuttgart 2014.*

110. H. P. Langtangen, *Finite Difference Computing with Exponential Decay Models.*

111. A. Tveito, G.T. Lines, *Computing Characterizations of Drugs for Ion Channels and Receptors Using Markov Models.*

112. B. Karazösen, M. Manguoğlu, M. Tezer-Sezgin, S. Göktepe, Ö. Uğur (eds.), *Numerical Mathematics and Advanced Applications - ENUMATH 2015.*

113. H.-J. Bungartz, P. Neumann, W.E. Nagel (eds.), *Software for Exascale Computing - SPPEXA 2013-2015.*

114. G.R. Barrenechea, F. Brezzi, A. Cangiani, E.H. Georgoulis (eds.), *Building Bridges: Connections and Challenges in Modern Approaches to Numerical Partial Differential Equations.*

115. M. Griebel, M.A. Schweitzer (eds.), *Meshfree Methods for Partial Differential Equations VIII.*

116. C.-O. Lee, X.-C. Cai, D.E. Keyes, H.H. Kim, A. Klawonn, E.-J. Park, O.B. Widlund (eds.), *Domain Decomposition Methods in Science and Engineering XXIII.*

117. T. Sakurai, S.-L. Zhang, T. Imamura, Y. Yamamoto, Y. Kuramashi, T. Hoshi (eds.), *Eigenvalue Problems: Algorithms, Software and Applications in Petascale Computing.* EPASA 2015, Tsukuba, Japan, September 2015.

118. T. Richter (ed.), *Fluid-structure Interactions.* Models, Analysis and Finite Elements.

119. M.L. Bittencourt, N.A. Dumont, J.S. Hesthaven (eds.), *Spectral and High Order Methods for Partial Differential Equations ICOSAHOM 2016.* Selected Papers from the ICOSAHOM Conference, June 27-July 1, 2016, Rio de Janeiro, Brazil.

120. Z. Huang, M. Stynes, Z. Zhang (eds.), *Boundary and Interior Layers, Computational and Asymptotic Methods BAIL 2016.*

121. S.P.A. Bordas, E.N. Burman, M.G. Larson, M.A. Olshanskii (eds.), *Geometrically Unfitted Finite Element Methods and Applications.* Proceedings of the UCL Workshop 2016.

122. A. Gerisch, R. Penta, J. Lang (eds.), *Multiscale Models in Mechano and Tumor Biology*. Modeling, Homogenization, and Applications.

123. J. Garcke, D. Pflüger, C.G. Webster, G. Zhang (eds.), *Sparse Grids and Applications - Miami 2016*.

124. M. Schäfer, M. Behr, M. Mehl, B. Wohlmuth (eds.), *Recent Advances in Computational Engineering*. Proceedings of the 4th International Conference on Computational Engineering (ICCE 2017) in Darmstadt.

For further information on these books please have a look at our mathematics catalogue at the following URL: `www.springer.com/series/3527`

Monographs in Computational Science and Engineering

1. J. Sundnes, G.T. Lines, X. Cai, B.F. Nielsen, K.-A. Mardal, A. Tveito, *Computing the Electrical Activity in the Heart.*

For further information on this book, please have a look at our mathematics catalogue at the following URL: www.springer.com/series/7417

Texts in Computational Science and Engineering

1. H. P. Langtangen, *Computational Partial Differential Equations.* Numerical Methods and Diffpack Programming. 2nd Edition
2. A. Quarteroni, F. Saleri, P. Gervasio, *Scientific Computing with MATLAB and Octave.* 4th Edition
3. H. P. Langtangen, *Python Scripting for Computational Science.* 3rd Edition
4. H. Gardner, G. Manduchi, *Design Patterns for e-Science.*
5. M. Griebel, S. Knapek, G. Zumbusch, *Numerical Simulation in Molecular Dynamics.*
6. H. P. Langtangen, *A Primer on Scientific Programming with Python.* 5th Edition
7. A. Tveito, H. P. Langtangen, B. F. Nielsen, X. Cai, *Elements of Scientific Computing.*
8. B. Gustafsson, *Fundamentals of Scientific Computing.*
9. M. Bader, *Space-Filling Curves.*
10. M. Larson, F. Bengzon, *The Finite Element Method: Theory, Implementation and Applications.*
11. W. Gander, M. Gander, F. Kwok, *Scientific Computing: An Introduction using Maple and MATLAB.*
12. P. Deuflhard, S. Röblitz, *A Guide to Numerical Modelling in Systems Biology.*
13. M. H. Holmes, *Introduction to Scientific Computing and Data Analysis.*
14. S. Linge, H. P. Langtangen, *Programming for Computations* - A Gentle Introduction to Numerical Simulations with MATLAB/Octave.
15. S. Linge, H. P. Langtangen, *Programming for Computations* - A Gentle Introduction to Numerical Simulations with Python.
16. H.P. Langtangen, S. Linge, *Finite Difference Computing with PDEs* - A Modern Software Approach.
17. B. Gustafsson, *Scientific Computing from a Historical Perspective.*
18. J. A. Trangenstein, *Scientific Computing.* Volume I - Linear and Nonlinear Equations.

19. J. A. Trangenstein, *Scientific Computing*. Volume II - Eigenvalues and Optimization.

20. J. A. Trangenstein, *Scientific Computing*. Volume III - Approximation and Integration.

For further information on these books please have a look at our mathematics catalogue at the following URL: www.springer.com/series/5151

Printed in the United States
By Bookmasters

Printed in the United States
By Bookmasters